# Emerging Infectious Diseases of the 21st Century

**Series Editor**
I. W. Fong, University of Toronto
Toronto, ON, Canada

Infectious diseases have been the scourge of mankind for centuries with such dreaded pestilence as the Black Death (plague) between the fourteenth and eighteenth centuries, and cholera epidemics intermittently from the sixteenth to the twentieth century. Previously dreaded diseases such as smallpox and poliomyelitis have been or will soon be eliminated from the world. However, many pathogens of the past are still with us today and are capable of creating havoc and pandemics (such as the Influenza virus). The advent of vaccines and antibiotics in the mid-twentieth century gave false hope to medical professionals and politicians that infectious diseases would be a relic of the past by the end of the twentieth century. Quite the contrary has occurred with new emerging infectious diseases being recognized, as we are currently witnessing with the AIDS pandemic, and the re-emergence of older, known microbes. The Emerging Infectious Diseases of the 21st Century series aims to address these new challenging infections and the surrounding issues facing physicians and mankind in this new century. Old or established pathogens such as Mycobacterium tuberculosis and malaria are now re-emerging or spreading across the globe in a more treatment-resistant form. These and other aggressive and difficult pathogens are addressed in the initial volume of the series, "Reemergence of Established Pathogens in the 21st Century" by world-renowned, leading experts in the field. Among the initial volumes is a review of the exciting area of new concepts of the relationship between microbes and the cardiovascular system: "Infection and the Cardiovascular System: New Perspectives" by a leading researcher in this area. Further volumes of the series address newly recognized pathogens such as Nipah virus, the emergence of old world pathogens in the new world, such as West-Nile virus, and many others.

I. W. Fong

# New Antimicrobials: For the Present and the Future

Springer

I. W. Fong
4th Fl CC Wing North, Rm 4179
St. Michael's Hospital
Toronto, ON, Canada

ISSN 1573-4684 ISSN 2627-5872 (electronic)
Emerging Infectious Diseases of the 21st Century
ISBN 978-3-031-26077-3 ISBN 978-3-031-26078-0 (eBook)
https://doi.org/10.1007/978-3-031-26078-0



This Springer imprint is published by the registered company Springer Nature Switzerland AG
The registered company address is: Gewerbestrasse 11, 6330 Cham, Switzerland

*Dedicated to my wife Cheryl*

# Preface

The alarming increase of microbial resistance to treatments is threatening the welfare of humans, animals, and the environments worldwide. Unless remedial actions are taken in a coordinated fashion, dire consequences will occur. The global population is facing an impending crisis in healthcare because of increasing antimicrobial resistance [AMR] and the inability to treat patients with severe infections due to lack of effective therapy. Estimates from 2013 in the United Sates indicated that AMR costs $55 billion per year, $20 billion for healthcare, and nearly $35 billion from loss of productivity. Presently, AMR infections caused 99,000 deaths annually in the United States alone. It has been projected that by 2050, that without effective therapies for AMR infections, 444 million people globally would succumb to infections and birthrates would decline rapidly as a result. Multiresistant microbes are not only a problem in hospital-associated bacterial infections, but also in tuberculosis, malaria, and increasingly in some viruses including human immunodeficiency virus [HIV], influenza, and cytomegalovirus [CMV] in cancer and transplant patients.

Development of new and innovative agents to combat the increasing and spreading antimicrobial resistance is paramount for our survival as a modern society. Thus, it is of importance to review new antimicrobials that have been approved in the last several years and their efficacy in meeting the challenges of emerging resistant microbes. Moreover, new drugs or novel agents in development to tackle the challenges of microbial resistance will be reviewed. This new edition *New Antimicrobials: For the Present and the Future* is a timely sequel to the recently published *Antimicrobial Resistance of the 21st Century*, Second Edition, and is an important addition to the current series, Emerging Infectious Diseases of the 21st Century. The first chapter reviews the current status of global antimicrobial resistance, mechanisms of antibiotic resistance, and measures in place to combat the exigency.

New antibiotics that have been approved in the last several years are reviewed in Chaps. 2, 3, 4, 5, 6, 7, and 8 and include new cephalosporins such as the novel siderophore, cefideocol, new β-lactam-β-lactamase inhibitors, new glycopeptides [telavancin, dalbavancin, and oritavancin], new oxazolininone [tedizolid], new tetracyclines [eravacycline and omadacycline], and the only new class of antibiotic [pleuromutilin] represented by lefamulin. Despite these new antibiotics, we are not close to conquering multidrug-resistant bacteria of concern, i.e., carbapenase-producing bacteria such as *Stenotrophomonas maltophilia, Acinetobacter* spp., and some strains *Pseudomonas aeruginosa.* We have more arsenal for treating multiresistant gram-positive bacteria such as methicillin-resistant *Staphylococcus aureus* and vancomycin-resistant *Enterococcus faecium.* Of major concern is the development of resistant bacteria to these new agents with limited use after a few years. Other disappointment include the fact that the new agents for multiresistant gram-negative bacteria have not shown superior efficacy than standard combination salvage therapy, but less toxicity and side effects.

New anti-tuberculous drugs reviewed in Chap. 9, bedaquiline, delamanid, and pretomanid, represent a significant advance in therapy for multidrug-resistant tuberculosis, but resistance to these agents has already been detected and development of further new agents will be needed.

The only new antifungal, isovuconazole, will not provide coverage for existing and emerging azole-resistant fungi, which are mainly seen in cancer centers and those performing stem cell transplantation, and the emerging multiresistant *Candida auris.* Similarly, there is lack of development of new antiparasitic agents. Chapter 11 reviews antiparasitic drugs recently approved in the United States for Chagas disease, malaria, fasciolosis, onchocerciasis, and African trypanosomiasis, but most of these agents were already used in endemic countries for years. The new agent for malaria, tafenoquine, is a derivative of primaquine, and the only advantage is the ease of a single dose for the radical cure of vivax malaria. Development of new antimalarial drugs to combat artemisinin-resistant falciparum malaria is sorely needed.

The greatest advance in antiviral therapy in the past decade was the development and marketing of agents for hepatitis C, reviewed in Chap. 15. Other new antiviral agents for HIV, cytomegalovirus, and influenza were also reviewed. Chapter 16 also reviews antiviral agents released for emergency use for COVID-19 infection.

The greatest and most urgent need is still development of new and novel antibiotics for multiresistant bacteria, but resistant strains of bacteria will always occur to every new agent. Thus, we have to keep one step ahead of these ubiquitous microbes and development of new antimicrobials will always be needed. Chapter 17 explores novel agents in development and non-antibiotic methods for treatment of infections.

Toronto, ON, Canada I. W. Fong

# Contents

## Part II New Systemic Antifungal Agents



## Part III Anti-Parasitic Agents

# Chapter 1
# Antimicrobial Resistance: A Crisis in the Making

## 1.1 Introduction

Antimicrobial resistance (AMR) is ancient and probably predates the evolution of Humanids to *Homo sapiens* or modern humans [1]. Although antibacterial resistance existed before the discovery of penicillin in 1928, it is largely due to overuse of antimicrobials in humans and animals. The current trend of increasing AMR threatens the antimicrobial effectiveness of increasing sphere of serious life-threatening infections due to bacteria, parasites, viruses, and fungi. Despite guidelines and antibiotic stewardship programs, antibiotic consumption from 2000 to 2015 in 76 countries had increased by 65%, and the global antibiotic consumption is projected to increase by 200% by 2030 [2]. There is an ever-increasing use of antimicrobials in livestock, as growth promoter and prophylaxis, since the practice was introduced in industrialized countries in 1950. In 2013, antimicrobial animal consumption globally was estimated to be 131,109 tons and is projected to reach 200,235 tons in 2030 [3]. Food animal production has plateaued in high-income countries since 2000 but has grown by 40–68% in low- and middle-income countries (LMIC) [4]. This has resulted in a corresponding increase in antimicrobial consumption by livestock in these countries. In Europe, regulations have been implemented to limit antimicrobial use in animal husbandry, while in the US consumer preference may have limited their use. A recent survey has found that AMR in animals is drastically rising in LMICs, with China and India representing the greatest hotspots of resistance and Brazil and Kenya are emerging hotspots [5]. The highest resistance rates are found in antimicrobials most commonly used in animals: tetracyclines, sulfonamides, penicillins, and quinolones. A recent report from the European Union on antimicrobial resistance from zoonotic indicator bacteria from animals and humans in 2016 found that resistance overall in critically important bacteria was generally uncommon, except for specific *Salmonella* serovars which showed very high multidrug resistant levels especially to ciprofloxacin and



I. W. Fong, *New Antimicrobials: For the Present and the Future*, Emerging Infectious Diseases of the 21st Century,
https://doi.org/10.1007/978-3-031-26078-0_1

**Table 1.1** Factors associated with antimicrobial resistance [AMR]

| Direct factors | Indirect factors |
|---|---|
| Overuse in healthcare | Poor sanitation |
| Overuse in farm animals | Poverty |
| Environmental contamination | Underdeveloped PHS |
| Easy access ["over-the-counter"] | Underuse of vaccines |
| Ineffective drugs | Low diagnostic methods |
| ↑Demand | ↑Prostheses and transplants |

Abbreviation: *PHS* public health system

extended-spectrum β-lactamase (ESBL) producers [6]. It is estimated that >75% of antimicrobials produced are used in food animals.

Low concentration of antimicrobials used in animal feed for growth promotion and mass prophylaxis promote the evolution of resistance, and food animal reservoir is a greater source of resistant genes than in humans. There is increasing evidence that antimicrobial resistance in animals can lead to resistant infections in humans [7–9]. While restricting use of antibiotics in food animals is associated with reduced antibiotic-resistant bacteria in animals and humans in direct contact with food-producing animals, the implication for the general population is less clear [10]. Thus, restricting use of antimicrobials in animals alone will not be sufficient to control AMR in humans [11]. However, the main driver of AMR globally is antibiotic pressure due to a combination of factors (see Table 1.1).

## 1.2 Antimicrobial Resistance: An Evolutionary Process

Soon after the discovery of each class of antibiotics, there would appear resistant bacteria with transmissible genetic elements or *r* genes which were considered a modern phenomenon. However, metagenomic analyses of ancient DNA from 30,000-year-old Beringian permafrost sediments identified diverse number of genes encoding resistance to β-lactams, tetracycline, and glycopeptide antibiotics [1]. Thus, AMR is a natural phenomenon which predates the discovery of antibiotics and is likely an evolutionary selective process for the survival of microbes living adjacent to antibiotic-producing bacteria or fungi (i.e., *Actinomycetes* and *Streptomycetes*). Antibiotic-producing *Actinomycetes* possess genes encoding resistance to the antimicrobials they generate and *Streptomyces* produce diverse β-lactamases, some of which may be responsible for clinical resistance [12, 13]. An environmental *Kluyvera* species appears to be origin of the CTX-M genes that encode the extended β-lactamase that hydrolyze third-generation cephalosporin [14]. It should not be surprising that antibiotic *r* genes and resistance-encoding integrons were found in the gut flora of isolated indigenous people who live in remote areas away from modern civilization without antibiotic exposure [15].

Antibiotic-resistant genomes are widespread in nature, and analysis of 13,293 genes yielded a core set of 4554 antibiotic resistant proteins/genes [16]. Functional metagenomic analysis of soil for bacterial resistance was reported to yield 2895 antibiotic resistance genes and represented all major resistance mechanisms [17]. However, recently 6000 antibiotic resistance genes were discovered in the bacteria from human gut [18]. Thus, humans harbor more microbial resistance genes than the environment.

## 1.3 Mechanisms of Microbial Resistance

AMR is a natural phenomenon that occurs over time through genetic changes of microbes, but this process is accelerated by high antimicrobial pressure due to overuse and misuse. Antimicrobial-resistant microbes are found worldwide in people, animals, food, and environment (soil and water), and transfer of resistance to humans can occur from any of these sources. Spread of antimicrobial resistance among humans is facilitated by poor infection control, inadequate sanitary conditions, and inappropriate food-handling. The ease of rapid modern transportation (air travel) has also facilitated the spread of antimicrobial-resistant microbes between peoples and animals of different countries across the world.

## 1.4 Bacterial Resistance

AMR is best studied and recognized in bacteria as antibiotics are the most frequently used antimicrobial agents in people and animals. Development and persistence of antibiotic-resistant bacteria are encouraged by the widespread use of antibiotics, broader-spectrum greater than narrow-spectrum agents, and longer-term use facilitate increased resistance more than shorter course. The widespread indiscriminate use of proton pump inhibitors (PPI) in healthcare facilities and by physician in general for gastric acid suppression also appears to be playing a role in intestinal colonization with multiresistant bacteria with possible cross-transmission in healthcare institutions [19].

Presently, there are >16 classes of antibiotics (used in the broad term) discovered, based on their structure and mode of action. Some are synthetic compounds (sulfonamides, quinolones, etc.) and others are natural antibiotics produced by microbes, most commonly from the phylum *Actinobacteria* of the genus *Streptomyces* (penicillin, streptomycin, etc.). The mechanisms of action of various antibiotics are important to review to appreciate the development and means by which bacteria develop AMR. Although there are seven different mechanisms by which antibiotics inhibit or kill bacteria, their actions result in the interruption of the synthesis and function of four main targets or pathways: (i) cell wall (beta-lactams, glycopeptides); (ii) cell membrane (polymixins, lipopeptides); (iii) nucleic acid

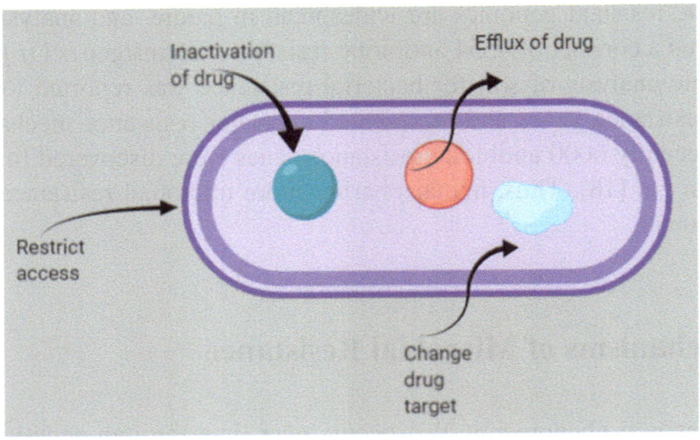


**Fig. 1.1** Mechanisms of bacterial resistance

synthesis (sulfonamides/pyrimidines [folate synthesis], quinolones [DNA gyrase], rifamycins [RNA polymerase]); and (iv) protein synthesis via 30S ribosomal subunit (aminoglycosides, tetracyclines) and 50S ribosomal subunit (macrolides, lincosamides, oxazolidinones, phenicols, streptogramins, and pleuromutilin) [20, 21].

Bacteria (and fungi) develop defense strategies to evade antibiotics by mutation or upregulation of existing resistosome broadly by four mechanisms (see Fig. 1.1): (i) restriction of access of these agents at the cell wall and membrane by changing entryways or limiting the number of entryways; (ii) changing the antibiotic target so the drug cannot fit or act at the target site or develop new cell processes that avoid using the antibiotic target; (iii) destroying or breaking down the drugs by enzymes; and (iv) extruding the agents from the cell by using efflux pumps. Table 1.2 shows an incomplete list of antibiotic classes and mechanisms of resistance [20–22].

### *1.4.1 Restriction of Access*

Gram-positive bacteria possess a thick complex but very permeable cell wall which readily allows antimicrobials and are easier to kill than gram-negative bacteria. However, resistance due to restricted penetration can occur as demonstrated by vancomycin-intermediate *Staphylococcus aureus* strains (VISA) that produce greatly thickened wall which decrease penetration and activity [23]. Gram-negative bacteria allow drug molecules diffusion through a bilayer of the outer membrane (OM) by porins. Small hydrophilic molecules (β-lactams and fluoroquinolones [FQ]) can cross the OM only through porins. Resistance to these classes of agents can occur through decrease in number of porin channels leading to decreased cell entry of the β-lactams and FQ. *Pseudomonas aeruginosa* acquire resistance to all classes of antibiotics through decreased OM permeability [20].

**Table 1.2** Classes of antibiotics and mechanisms of resistance

| Antibiotic class | Resistance mechanisms | New addition |
|---|---|---|
| Aminoglycoside | Decrease uptake, enzymatic modification, and efflux | Plazomicin |
| Carbapenems | Carbapenemase, changed porin selectivity, efflux | Mero.-Vaborbactam Imipen.-Relebactam |
| Cephalosporins | Cephalosporinase, porin selectivity, efflux | Ceftaz.-Avibactam Cefiderocol |
| Penicillins | Penicillinase, altered PBPs, efflux | |
| Monobactams (Aztreonam) | Efflux, altered PBP? | |
| Cationic peptides (Colistin) | Efflux, altered target | |
| Fidaxomicin (macrocyclic lactone) | Rare, unclear, *rpo* [B] mutation | |
| Fosfomycin | Decrease uptake, altered target, and degradation | |
| Fusidic acid | Decrease uptake, altered target | |
| Glycopeptides | Decrease uptake, modified target. | Oritavancin |
| Lipopeptide [Daptomycin] | Modified net cell surface charge, altered target | |
| Lincosamide [Clindamycin] | Altered target, efflux, nucleotidylation | |
| Macrolides | Altered target [methylation], efflux | |
| Metronidazole | Decrease uptake, efflux | |
| Nitrofurans | Unclear—efflux? altered target? | |
| Oxazolidinones (Linezolid) | Altered target, efflux | Torezolid |
| Phenicols (Chloramphenicol) | Acetylation, altered target, efflux | |
| Pleuromutilin | Unclear [altered target?] | Lefamulin |
| Quinolones | Altered target, efflux, acetylation of drug | |
| Sulfonamides/ Trimethoprim | Altered target, decrease uptake, efflux | |
| Tetracyclines | Decreased uptake, altered target, efflux, | Omadacycline, Eravacycline |

### *1.4.2 Modification of Target*

Modification of target molecules by natural variation or acquired changes in the target sites of antimicrobials is a common mechanism of drug resistance. Spontaneous mutation of a bacterial gene on the chromosome often results in drug target sites modification. Drug interaction with the target molecule is usually very specific and minor alterations can affect antibiotic binding and decrease action. Examples of drug target modifications can be subdivided as follows: (a) Alterations in the 30S subunit can result in resistance to tetracyclines and aminoglycosides (AG) and to the 50S subunit lead to resistance to macrolides, chloramphenicol, lincosamides, and streptogramin B [24]. (b) Modification of the penicillin-binding proteins (PBP) is a common mechanism used by gram-positive bacteria to reduce affinity to β-lactam drugs. This is demonstrated by mutation in the PBP leading to *Enterococcus faecium* resistance to ampicillin and *Streptococcus pneumoniae* to penicillin. *S. aureus* resistance to methicillin/oxacillin is through mec gene A that encodes PBP2a protein with reduced affinity to the β-lactams leading to methicillin-resistant *S. aureus* (MRSA). The mec A gene is transmitted through a large mobile genetic element, "staphylococcal cassette chromosome mec," that is integrated into the chromosome of MRSA [25]. There is resistance to all β-lactam agents, and cross-resistance to macrolides, clindamycin, aminoglycosides, and less commonly tetracyclines may be seen. (c) Cell wall precursor modification (i.e., D-alanyl-alanine changed to D-alanyl-lactate) will lead to glycopeptide resistance by preventing their binding to D-analyl-D-alanine residues of the peptidoglycan precursors. Van A type resistance leads to high resistance of *E. faecium* and *E. faecalis* to vancomycin and teicoplanin, whereas Van B and Van C type resistance show resistance to vancomycin but sensitive to teicoplanin [26]. (d) Quinolones bind to DNA gyrase A subunit and mutated DNA gyrase and topoisomerase IV leads to FQ resistance. The resistance mechanism involves the modification of two enzymes: DNA gyrase (coded by genes gyr A and gyr B) and topoisomerase IV (coded by genes par C and par E), and mutation in genes gyr A and par C leads to failure of FQ to bind to the target site [27]. (e) Ribosomal protection mechanisms imparting resistance to tetracyclines. (f) RNA polymerase mutation conferring resistance to rifampin [20].

## 1.4.3 Degradation by Enzymes

Antibiotic degradation or modification by bacterial enzymes is one of the most commonly recognized mechanisms of AMR. This mechanism for self-defense by the antibiotic-producing microbe was recognized in 1970 in soil bacteria of the genus *Streptomyces* [28]. This mechanism is frequently used by gram-negative bacilli and to a lesser degree by gram-positive bacteria. The three main groups of enzymes that inactivate antibiotics are (1) β-lactamases, (2) aminoglycoside-modifying enzymes, and (3) chloramphenicol acetyltransferases (AAC).

### 1.4.3.1 Beta-lactamases

There are >900 β-lactamases circulating and identified in bacteria to date. β-lactamases hydrolyze nearly all β-lactam agents that have ester and amide bond, i.e., penicillins, cephalosporins, monobactams, and carbapenems. The β-lactamases can be classified into four groups (Ambler structural system): Class A β-lactamases (referred to as penicillinase was the first β-lactamase discovered in 1940) include the penicillinase produced by *S. aureus* and the *Enterobacteriaceae,* termed TEM-1, TEM-2, and SHV-1 which have no activity against the cephalosporins (especially expanded spectrum) [20]. TEM-1 is the most common β-lactamase found in gram-negative bacteria, accounting for ampicillin resistance *in Escherichia coli, Klebsiella pneumoniae, Haemophilus influenzae*, and *Neisseria gonorrhoeae.* Mutations in the *Enterobacteriaceae* gave rise to the extended spectrum β-lactamases (ESBLs) that provide multi-resistance to penicillins, cephalosporins, and cephamycins, but the carbapenems are usually effective. CTX-M β-lactamases also belong to Class A, they are mainly found in *Salmonella enterica* serovar *Typhimurium* and *E. coli*, which acquire plasmid β-lactamase genes normally found on commensal bacteria and produce hydrolysis of cefotaxime more than ceftriaxone, ceftazidime, and cefepime due to structural differences [Wikipedia, beta-lactamases, 3/26/2020]. Class B β-lactamases are the metallo-β-lactamases (MBL), containing zinc ions, that can hydrolyze nearly all β-lactam drugs, and unlike other classes (A, C, and D enzymes), they are resistant to the β-lactamase inhibitors, i.e., clavulanic acid, sulbactam, tazobactam, and avibactam, and carbapenems [29]. These enzymes are divided into three subclasses based on the zinc content, but the most relevant include VIMs (Verona integron-encoded MBL), IMPs (imipenases), and NDMs (New Delhi MBL). Class C β-lactamases (called cephalosporinases) hydrolyze all cephalosporins and other β-lactams except carbapenems; the best known is Amp C β-lactamase which is common in ESBL bacteria [20]. Class D β-lactamases (OXA) are oxacillin-hydrolyzing enzymes (weakly inhibited by clavulanic acid) which are most commonly found in *Pseudomonas aeruginosa* and the *Enterobacteriaceae*. The OXA type can result in the ESBL phenotype, and some of the enzyme can hydrolyze cefotaxime, cefepime, and ceftazidime [Wikipedia].

ESBL-producing gram-negative bacteria infections have been a challenge to treat in hospitalized and chronic care facilities worldwide for the past two decades [30]. ESBLs are transmissible (plasmid mediated) β-lactamases that hydrolyze extended-spectrum cephalosporins with oxyimino side chains, i.e., cefotaxime, ceftriaxone, ceftazidime, and aztreonam [Wikipedia]. The plasmids encoding ESBL frequently carry genes encoding resistance to other drug classes (aminoglycoside, quinolone, etc.). Although the carbapenems are considered treatment of choice for severe infection by ESBL-gram-negative bacilli, carbapenem-resistant (primarily ertapenem-resistant) isolates have been reported. The ESBLs were primarily derived from genes for TEM-1, TEM-2, and SHV-1 by mutations, but subsequently these enzymes include other classes of β-lactamases.

#### 1.4.3.2 Aminoglycoside-Modifying Enzymes

Aminoglycoside-modifying enzymes (AME) are the most common mechanism of resistance to this class of antibiotics. There are over 100 AME which can be divided in three subclasses: aminoglycoside [A]-acetyltransferases (AACs), A-nucleotidyltransferases (ANTs), and A-phosphotransferases (APHs) [31]. These enzymes reduce the affinity of modified agents and impair binding to the 30S ribosomal subunit, resulting in resistance to aminoglycosides and quinolones [20]. AME are identified in gram-negative bacilli, *Mycobacterium tuberculosis*, *S. aureus*, *E. faecalis*, and *S. pneumoniae* [20].

#### 1.4.3.3 Chloramphenicol Acetyltransferase

Chloramphenicol resistance in gram-positive and gram-negative bacteria, including *H. influenza*, is most common through modification of the antibiotic by acetyltransferases. The modified antibiotic is unable to bind to the ribosomal 50S subunit [26].

#### 1.4.3.4 Efflux Pumps

Although efflux pump was first described as a mechanism of tetracycline resistance in *E. coli* in 1980 [32], it is now recognized as an ancient evolutionary protective process that constitutes the most ubiquitous system present in all organisms, including bacteria, eukaryotic pathogens such as *C. albicans* and *P. falciparum,* etc., but also mammals including human cells [33, 34]. Essentially efflux pumps are MDR resistant mechanisms present in all microorganisms. They are nearly always chromosomally encoded, conserved at the genetic and protein level, and most bacterial strains of the same species have the same chromosomally coded efflux pumps [35]. MDR efflux pumps are present in all organisms, but are tightly regulated and low-moderate expression may result in intrinsic resistance (i.e., *Ps. aeruginosa*), but acquired resistance may occur in two ways. In chronic infections, antibiotic pressure may cause overexpression of MDR efflux pumps due to mutations in the genes that control downregulation of their expression; phenotypic resistance occurs transiently from the presence of specific inducers of the efflux pumps expression [35]. The efflux systems can actively extrude a variety of compounds besides antimicrobials, such as heavy metals, toxins, organic solvents, dyes, detergents, and others. Overexpression of a single efflux pump can give resistance to multiple antimicrobials, but simultaneous overexpression of multiple efflux pumps may occur with some organisms [35].

### *1.4.4 Genetic Mechanisms of Bacterial Resistance*

AMR can either be intrinsic, adaptive, or acquired. Intrinsic antibiotic resistance is common in the environmental bacteria, and the mechanisms are normally chromosome-encoded, including nonspecific efflux pumps, inactivating enzymes, and permeability barriers [28]. These mechanisms are fixed in the core genetic makeup of the microbe and often confer low level resistance in the original host. Normal commensal flora and environmental bacteria with intrinsic mechanisms of resistance can become opportunistic pathogens in immunocompromised hosts [31]. Adaptive antibiotic resistance of bacteria occurs as a result of harmful environmental exposure (changes in nutrients or subinhibitory concentration of antibiotics) that results in transient changes in gene and protein expression with tolerance to the antimicrobial [36]. Acquired antibiotic resistance occurs by acquisition of exogenous genes from other bacteria by transduction of free DNA by bacteriophages, or conjugation via plasmids, or through mutation of existing genes. Dissemination of resistant genes by plasmids is considered the most prevalent means among various bacterial species. Transfer of resistant genes by plasmids between bacteria is most expedient in high-density settings such as the gut of humans or animals, biofilms, hospitals, and conditions with co-infection [37]. This process is facilitated by transposons and integrons incorporated in plasmids or phages for conjugation [28]. Transposon is a DNA sequence than can change its position within a genome ("jumping genes") and can carry resistance genes from plasmid to plasmids or from a DNA chromosome to plasmid or vice versa. Integron is a mobile DNA element that can capture and carry genes (expression or gene cassettes encoding antibiotic resistance), by site-specific recombination. Plasmid is an extrachromosomal self-replicating, double-stranded DNA molecule that carries genes not essential for cell growth, such as antibiotic resistant genes, that can be transferred from cell to cell by conjugation or transduction [Dorland Medical Dictionary].

It is now evident that the environment is an important source for pathogenic bacteria to acquire antibiotic-resistant genes. This process may involve four stages: (i) emergence of novel resistance genes, (ii) mobilization (transposons/integrons), (iii) transfer to pathogens (by plasmids), and (iv) dissemination by horizontal transfer [28]. Novel resistance genes are likely occurring all the time in the environment and the most important factor to promote persistence of the resistance genes is selective pressure. The predominant source of selective pressure is the widespread and indiscriminate use of antibiotics, which leads to dominance of resistant and multiresistant strains of bacteria among human pathogens, in the environment near human activities (i.e., antibiotic manufacturing plants), and in food animal farms. It has been estimated that in the past 50 years, millions of metric tons of antibiotic compounds have been released in the biosphere [15], which is undoubtedly contributing to resistant genes in the environment.

Environmental sampling studies have revealed multiresistant *r* genes to 7–8 antibiotics, which has been labeled environmental antibiotic "resistome" [38]. Moreover, many environmental bacteria can subsist and grow on 18 different antibiotics as the sole source of carbon and nitrogen, called "subsistome," including aminoglycoside, FQ, and others [39]. Most of strains identified were proteobacteria, >40% are *Burkholderia* spp., and pseudomonads were also represented.

The origin of antimicrobial resistance and generation of *r* genes for horizontal spread is through the process of "natural selection" in which evolutionary change occurs through genetic mutation. In vitro resistant mutants can be generated spontaneously to virtually any antibiotics, but the frequencies vary markedly depending on the agent and microbial species, with most frequencies usually $\leq 10^{-6}$ [40]. Resistant mutants may be less fit than wild-type organisms, but compensatory mutations may occur so the resistant mutants become equally fit as the wild-type organisms and some strains even maintain the resistant mutation in the absence of the antibiotic selective pressure [41]. Bacterial resistance to some classes of antibiotics occurs primarily by genetic mutation rather than by acquisition of *r* genes by horizontal transfer, i.e., by plasmids. Resistant mutations readily occur to rifamycins, fusidic acid, and streptomycin when used as monotherapy and less readily to FQ and oxazolidinones (linezolid) [40]. Unlike rifampin, resistance to linezolid is extremely rare clinically and in vitro to generate, as a single mutation in one gene is insufficient to confer phenotypic resistance. In some bacterial species, resistance occurs primarily or solely by genetic mutation, such as *Mycobacterium tuberculosis* and *Helicobacter pylori*. *M. tuberculosis* develops resistance to all anti-tuberculosis agents by mutation, thus the need for multidrug therapy. Similarly, *H. pylori* requires at least two antibiotics and PPI to avoid resistant mutation, as chromosomal mutation is responsible for resistance to clarithromycin (in 23S rRNA), amoxicillin (changes in penicillin-binding protein 1), metronidazole (multiple genes), and tetracycline (in 16S rRNA and other genes) [40]. In addition, many *Enterobacteriaceae* carry chromosomally encoded cephalosporinases resulting in resistance to broad-spectrum β-lactam agents. *Ps. aeruginosa* which is intrinsically resistant to many antibiotics, in certain environment (cystic fibrosis, bronchiectasis), is difficult to eradicate because of biofilm existence and mutations that result in overexpression of many intrinsic efflux pumps [42]. In this setting, the pseudomonas represent a hypermutator strain of bacteria, which have increased mutation rates in genes relating to DNA repair and replication constancy [42]. In the *Enterobacteriaceae* members that produce ESBL (150 TEM variants and 90 SHV variants), some variants have undergone mutations which render the enzymes capable of hydrolyzing extended spectrum cephalosporins or able to resist the action of β-lactamase inhibitors [43]. Moreover, mutation is necessary for the development and acquisition of new resistant genes in the TEM family of β-lactamases.

Conjugation, transfer of DNA via cell surface pili or adhesions, is considered the most important means for bacteria to disseminate antibiotic-resistant genes usually utilizing transposons, integrons, and plasmids. The worldwide spread of resistant genes to many drug classes is attributed to the transfer of plasmids in pathogens with antibiotic-resistant genes encoding resistance to β-lactams, tetracyclines, sulfonamides, quinolones, aminoglycosides, and many others [44]. The increasing reports of pathogens harboring plasmids for carbapenem resistance and the spread of plasmid-encoding colistin resistance to many continents are of major concern [45]. Plasmids are equivalent to a carriage basket, as multiple antibiotic-resistant genes can be co-localized on the same plasmid allowing for the spread of multidrug resistance. The spread of pan-resistant *Enterobacteriaceae* is now a reality.

The horizontal spread of resistance genes by transformation and transduction are considered less important than conjugation in dissemination, but understanding all the means of gene transfer to pathogens and mechanisms of spread of antibiotic resistance is necessary for their control. Transformation is the process by which certain bacteria (first demonstrated with *S. pneumoniae*) are capable of uptake, integration, and functional expression of naked fragments of extracellular DNA [46]. Bacteria could use this mechanism to evade antibiotics by exchanging resistant genes. Intra- and inter-species of DNA could be transferred by this means under certain conditions: presence of extracellular DNA in the environment; the recipient bacteria must be in a state of competence; and the translocated DNA must be stabilized into the recipient genome. Exposure to antibiotics can induce competence in many species of bacteria and enhance transformation of resistant genes. In addition, it has been shown that natural transformation facilitates the transfer of transposons, integrons, and gene cassettes (which may contain antibiotic-resistant genes) between bacterial species [47].

Transduction is the process by which bacteriophages transfer genes that are advantageous to the microbial host but also promotes their survival and dissemination. The genetic materials that can be transferred include chromosomal DNA, plasmids, transposons, and genomic islands [48]. Bacteriophages have been documented to transfer antibiotic resistance genes for many antibiotics (erythromycin, tetracycline, β-lactams, and aminoglycoside) by various bacterial species: streptococci, enterococci, *E. coli*, *Salmonella*, and *S. aureus* (MRSA) [46]. It is now evident from recent studies using metagenomic methods on various environmental samples (including wastewater from hospitals), patient samples (feces, respiratory secretion), and animals (feces, meat) that bacteriophages are significant reservoirs of many antibiotic resistant genes and are capable of transducing resistance genes to diversified bacterial communities.

## 1.5 The Toll of Antimicrobial Resistance

AMR presents an urgent threat to the public health systems of the world. The consequences can be measured by its effects on human lives and function, the healthcare system, and the economic burden. With respect to patient outcome, there is increasing evidence that AMR infections increase the mortality and morbidity of affected subjects. Compared to patients with infection due to non-resistant bacteria, those infected with AMR bacteria have double the risk of serious complication and triple the risk of death [49]. Presently, it is estimated that about 700,000 people lose their lives as a result of AMR infection worldwide each year [49], which may escalate to ten million by 2050 (see Table 1.3). In the United States (US) in 2013, the Centers for Disease Control and Prevention (CDC) estimated that two million persons were infected with AMR pathogens each year and 23,000 patients died as result annually [50]. About two thirds of those deaths were associated with infections caused by multidrug-resistant (MDR) pathogens. MRSA alone appears to be

**Table 1.3** Projected mortality rates for various conditions by 2050 worldwide

| Conditions | Deaths annually |
|---|---|
| Cancer | 8.2 million |
| Cholera | 100,000–120,000 |
| Diabetes | 1.5 million |
| Diarrheal disease | 1.4 million |
| Measles | 130,000 |
| Tetanus | 60,000 |
| Road traffic accidents | 1.2 million |
| Antimicrobial resistance | 10 million (presently 700,000) |

causing nearly 50,000 deaths yearly in the US and Europe combined [49]. Recently, the trend of MDR bacterial infections in US hospitalized patients from 2012 to 2017 has been reported by the CDC [51]. These patients accounted for 41.6 million hospitalization (>20% annually). In 2017, MDR bacteria accounted for 622,390 infections of which (surprisingly) 83% had their onset in the community. Although between 2012 and 2017, the incidence decreased for MRSA, vancomycin-resistant enterococcus (VRE), and MDR *Ps. aeruginosa* infection, the incidence of carbapenem-resistant *Enterobacteriaceae* did not change and the incidence of ESBL infection increased by 53% (mainly from community-onset cases) [51].

MDR-TB is a global security risk and public health crisis. In 2018, the WHO estimated there were 484,000 new cases with rifampin resistance, of which 78% had MDR-TB and 6.2% of these were extensively drug-resistant (XDR-TB) [WHO, Tuberculosis, 2020]. Moreover, only 56% of MDR-TB patients are presently treated successfully. Although in the past decade, there has been a decline in the global incidence of malaria, 228 million cases with 405,000 deaths from the infection were reported in 2018 [WHO, Malaria Report 2019]. The trend of increasing artemisinin resistance *in Plasmodium falciparum* in Southeast Asia (SEA) is causing high treatment failures with artemisinin combination therapy (ACT) in Cambodia, Vietnam, Thailand, Laos, and Myanmar [52]. In addition, there is evidence of increasing *Plasmodium vivax* resistance to chloroquine in SEA, up to 10% in Indonesia [WHO, Malaria Report 2019]. Hence, the prospect of global elimination of malaria in this century appears bleak. The WHO has listed microbes or infections which are of concern for AMR that need close surveillance and coordinated global action (see Fig. 1.2 for priority areas for development of new agents).

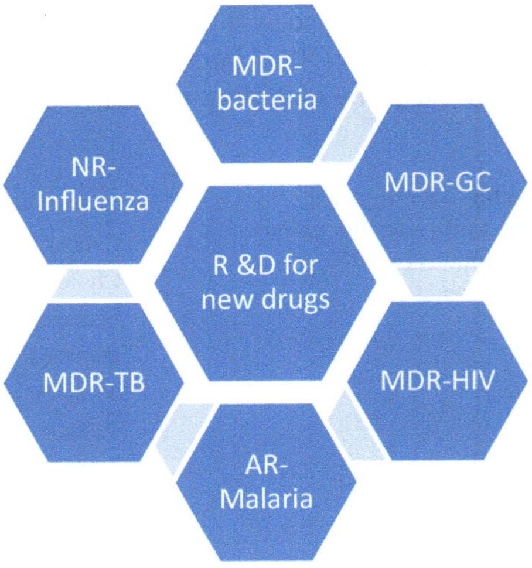


**Fig. 1.2** Priority infections for research and development of new antimicrobials. Abbreviations: *AR* artemisinin resistant, *GC* gonorrhea, *MDR* multidrug-resistant, *HIV* human immunodeficiency virus, *NR* neuraminase-resistant, *R & D* research and development, *TB* tuberculosis

### 1.5.1 *Effect on Healthcare*

With respect to the healthcare perspective, AMR is having enormous effect on healthcare costs and public health, the burden being felt more in low- and middle-income than high-income countries. Calculation of the cost to healthcare systems is complex and multifaceted including expensive second-line drugs (often with increased side effects), cost of isolation/containment, additional diagnostics, more intensive care (ICU) utilization, cost of surveillance, cancelling of elective surgeries, closure of some units (i.e., dialysis, chemotherapy, etc.), longer hospital stay and turnover, and decreased revenues. The additional cost varies in studies but could be more than $2 billion every year, and by 2050 the annual cost globally has been projected to vary from $300 billion to more than $1 trillion [53, 54]. Just enumerating the economic cost of five drug-resistant pathogens (*S. aureus* [MRSA], *E. coli*, *K. pneumonia*, *A. baumannii*, and *Ps. aeruginosa*), narrowly defined as incremental cost and indirect productivity losses, the annual cost is estimated to be $0.5 billion and $2.9 billion in Thailand and the US, respectively [55]. Despite measures to contain overuse of antibiotics, including antibiotic stewardship programs in hospital in the US, treatment of AMR infections had doubled since 2002, exceeding $2 billion annually [56]. Treatment of a patient with a MDR infection cost the hospital (on average) an additional $10,000–40,000 (US) compared to one with a sensitive organism causing infection. The human toll from AMR infections is quite substantial. In the US alone, antibiotic-resistant bacterial hospital infections result in 99,000 deaths yearly, and by 2050 it is projected that without a solution to the current trend in AMR, up to 444 million people worldwide would die from infections, resulting in rapid decline in birth rates [57].

### 1.5.2 *Economic Effect*

On a broader economic scale, the World Bank research indicate that AMR is a threat to our future and would increase the rate of poverty with greater impact on low-income countries than others [58]. Previous estimates by CDC from 2013 indicated that AMR cost $55 billion per year in the US, $20 billion for healthcare and nearly $35 billion from loss of productivity [59]. More recent studies show that the annual global gross domestic product (GDP) could shrink by about 1% with loss of 5–7% in developing countries by 2050 ($100–210 trillion) [60–62]. MDR-TB alone could cost the world $16.7 trillion by 2050 according to some estimates [61]. The economic impact of AMR is more complex than shrinkage of the GDP. Labor shortage from sickness and premature deaths is predicted to occur in ten years at the current level of AMR, resulting in decrease in global exports from labor-intensive sectors especially from Eurasia by 2050 (according to the World Bank). The impact of AMR will be realized as well by the livestock industry due to sickness and mortality of food animals, resulting in shortages of meat and dairy products, with persistent trend of AMR producing a 11% loss in livestock by 2050 [62]. The loss in food animal production will affect export and trade with decline in gross national product, decrease employment, average income, and economic stagnation or decline [63].

## 1.6 Global Response to Antimicrobial Resistance

The Global Action Plan on Antimicrobial Resistance was developed in 2015 by the WHO, the Food and Agriculture Organization of the Unite Nations (FAO), and the World Organization for Animal Health (WOAH), recognizing the high level of antimicrobial resistance from inappropriate use of these drugs in humans, animals, food, agriculture, and aquaculture farms. World leaders from 193 countries agreed to address the spread of AMR at a high-level meeting at the 71st UN General Assembly in September 2016 [Global Health, JAMA 2016; 316: 1936]. The "Global action plan on antimicrobial resistance" 5 strategic objectives were to (i) improve awareness and understanding of antimicrobial resistance; (ii) strengthen surveillance and research; (iii) reduce incidence of infection; (iv) optimize the use of antimicrobial drugs; and (v) ensure sustainable investment in countering antimicrobial resistance. To address these issues, the WHO initiated a broad, coordinated approach with the UN Member States to correct the root causes of AMR across multiple sectors, in particular human health, animal health, and agriculture. The WHO provided support to Member States to develop national action plans on AMR based on the global action plan. These initiatives included the following: World Antibiotic Awareness Week (a global, multiyear campaign); the Global AMR and Use Surveillance System (GLASS), the WHO supports a standardized approach to the collection, analysis, and sharing of data related to AMR at a global level to inform decision-making and actuate local, national, and regional action; Global Antibiotic Research and Development Partnership (GARDP), to encourage research and development through public-private partnerships for producing novel and new antimicrobials; and Interagency Coordination Group on AMR (IACG), to improve coordination between international organizations and to ensure effective global action against the threat of AMR [WHO, Antibiotic resistance, February 2018]. Figure 1.3 outlines the key factors in tackling AMR.

What progress has been made since 2016 in tackling drug-resistant infections globally? A recent review of the progress made by the Global Health Program on Antimicrobial Resistance was reported in Oct. 2019 [Charles Clift, Chatham House]. The findings of the review are summarized as follows:

- Little progress has been made in transforming research and development incentives for antibiotics, vaccines, and diagnostics.
- Significant advances in reducing antibiotic use in agriculture in high-income countries, but much more needed to convince low- and middle-income countries [LMICs] to reduce antibiotics in this area.
- Although there has been greater investment in awareness raising, questions remain on the impact and effectiveness in changing behavior.
- Restriction of over-the-counter antibiotics is yet to be implemented in LMICs, which is hampered by poor living conditions and access to healthcare.
- Limitations in LMICs that affect infection control and antibiotic overuse awareness messages are high rates of unhygienic conditions in the community and healthcare facilities.

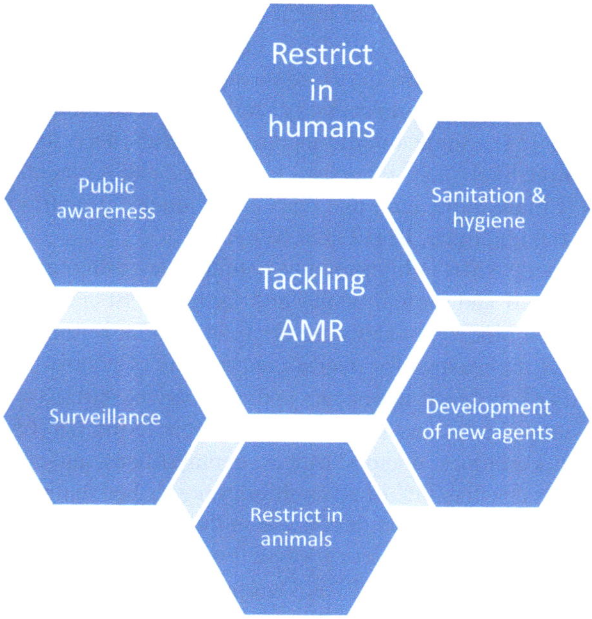


**Fig. 1.3** Outline of the key factors in tackling antimicrobial resistance (AMR). Other elements not shown include vaccines and alternatives, rapid diagnostics, Global Innovation Fund, and International Coalition for Action

- Providing quality healthcare to all and moving toward universal health coverage in LMICs are crucial in addressing the problems of both adequate access to antibiotics and restriction of over-the-counter sales.
- Funding agencies (World Bank, International Monetary Fund [IMF]) and governments should put greater emphasis in investments in providing clean water, sanitation, and housing to reduce the reliance on antibiotics in the long term in LMICs.
- Although investments have been made in improving surveillance of antibiotic use and resistance, particularly for humans, more effort is required to create surveillance systems that provide data sufficiently accurate to influence policy and actions. This applies also to antibiotics and resistant genes circulating in the environment.

## *1.6.1 Comments on Global Response*

The approach to tackling the AMR crisis is a difficult one and should be different for LMICs compared to high-income countries where the root causes may be different, but overlap exists. The funding agencies and governments of high-income countries should invest and assist lower-income countries to improve their public health systems and healthcare facilities. Proper housing and clean running water with proper hygienic conditions are much needed for thousands of communities worldwide, and these measures with basic childhood vaccines (measles, mumps, pertussis, conjugate *S. pneumoniae, H. influenzae,* and rotavirus) could save hundreds of millions of lives and unnecessary antibiotics with reduction of antibiotic-resistant bacteria [64]. Some middle-income countries need assistance in these areas, but they more likely need incentives to decrease environmental pollution with antimicrobials (China, India, etc.) and discourage the use of antibiotics in farming and agriculture (China, India, Brazil, and Kenya).

The forces driving AMR in high-income countries are different and are two-fold, healthcare associated and animal health and farming associated, including aqua-care. The use of antibiotics in healthcare are in outpatient settings, nursing homes, and hospitals, and the reasons for overuse and means of control may be different. The vast majority of antibiotic use occurs in the outpatient setting (≈80%) and > 250 million outpatient prescriptions are written every year in the US [65]. These outpatient settings include clinics, doctors' offices, dental offices, and emergency rooms, and the CDC estimates that 30% of all antibiotics prescribed in the outpatient clinics are unnecessary [CDC, 2017, Antibiotic use in the United States]. However, over-prescribing of antibiotics may be even worse than this estimate. In a recent unpublished study presented at IDWeek 2018, October 5 [Infectious Diseases Society of America] [JA Linder], of >500,000 antibiotic prescriptions, nearly half of the time antibiotics were prescribed without an infection-related diagnosis and one in five prescriptions were provided without an in-person visit. Antibiotic stewardship programs for outpatient management may be of value for ERs and outpatient clinics but would be difficult to implement for doctors' offices outside hospitals, especially primary care physicians. Most inappropriate outpatient antibiotic prescribing is for viral respiratory infections, i.e., viral bronchitis, otitis, and sinusitis, and secondly for unnecessary broad-spectrum agents. In a recent study, clinical education on antibiotic use resulted in >50% reduction in inappropriate antibiotic prescribing for acute respiratory infections at 1 year, but behavior interventions did not have sustained effect except for monthly emails with peer comparison practices [66].

Patients in nursing homes are commonly prescribed antibiotics for urinary tract, respiratory tract, and skin and soft tissue (pressure ulcers) infections. Some studies have reported that two thirds of patients in nursing homes are prescribed antibiotics each year and up to 75% may be inappropriate [67, 68]. In a small study of 9 nursing homes, the CDC found 11% of residents were receiving antibiotics and 40% lacked prescribing information [CDC, 2017]. Hence, a larger study was implemented.

Antibiotic overuse and overprescribing in acute care hospitals were recognized as major factors leading to multiresistant bacterial infections associated with hospital-acquired infections, resulting in increased demand for broader, more expensive, or more toxic agents which propagated the spiral or vicious circle of greater microbial resistance. The CDC estimated that 70% of the two million infections acquired in US hospitals each year are resistant to at least one commonly used antibiotic, and 20–50% of antibiotics prescribed in acute care hospitals are unnecessary or inappropriate [69]. Antibiotic stewardship programs were introduced in hospitals just over 30 years ago to improve inappropriate antibiotic prescribing and decrease antibiotic resistance rates [70, 71]. These programs are now widely implemented in hospitals of Europe and North America, and recent reviews of randomized and non-randomized studies have confirmed their value [72, 73]. Two types of antibiotic prescribing interventions are usually employed: restrictive, limit on which antibiotics can be prescribed, and enablement technique, education, verbal and written reminders, evaluation, audit, and feedback of individual physicians prescribing habits, but sometimes the combination of the two interventions. Antibiotic overuse or inappropriate use are usually considered when not clinically indicated, use of broad-spectrum when narrow-spectrum agents are more suitable, failure to transition from parenteral to oral therapy, and excessive duration than necessary. In a recent multihospital cohort study from Michigan, assessing excessive duration of antibiotic for pneumonia in hospitalized patients, two thirds (67.8%) were prescribed excess antibiotic therapy (mainly prescribed at discharge) and each excess day of treatment was associated with 5% increase in adverse events [74].

Even in industrialized countries, restriction of antimicrobials in agriculture and farming for growth promotion or prophylaxis is not uniform. The European Union ban antimicrobial growth promoters in 2006, while Australia and New Zealand instituted partial ban, and the US restraint is voluntary [75]. In 2014, the Canadian government restricted the use of growth-promoting antibiotics in livestock, but farmers were able to bypass this ban by importing and stocking large supplies of antimicrobials without a prescription. This loophole was closed in December 2018 when farmers were required to obtain a veterinary prescription for antibiotics in livestock [Nicole Williams/Canadian Broadcasting Corporation]. Hence, in high-income countries, much improvement is still needed to limit the overuse of antimicrobials in humans and farm animals.

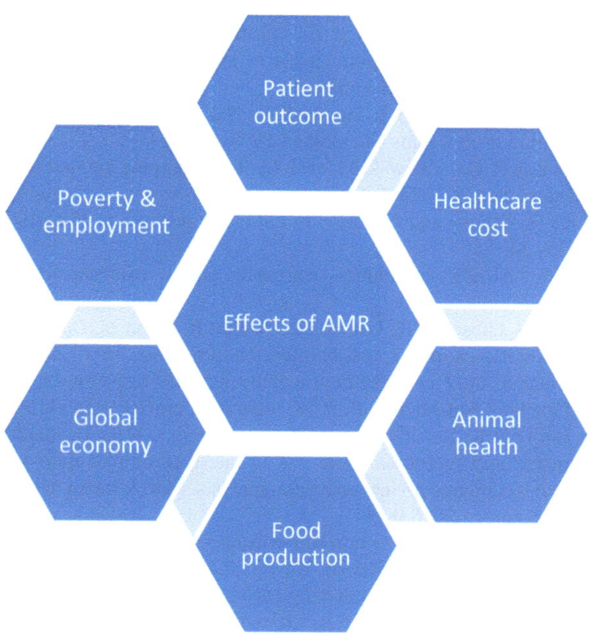

## References

1. D'Costa VM, King CE, Kalan L, et al. Antibiotic resistance is ancient. Nature. 2011;477:457–61.
2. Klein EY, Van Boeckel TP, Martinez EM, et al. Global increase and geographic convergence in antibiotic consumption between 2000 and 2015. Proc Natl Acad Sci USA. 2018;115:E3463–70.
3. Van Boeckel TP, Browser C, Gilbert M, et al. Global trends in antimicrobial use in food animal. Proc Natl Acad Sci U S A. 2015;112:5649–54.
4. Food and Agriculture Organization of the United Nation. (2017). http://www.fao.org/faostat/en/#home
5. Van Boeckel TP, Pires J, Silvester R, et al. Global trends in antimicrobial resistance in animals in low-and middle-income countries. Science. 2019;365:eaaw1944.
6. European Food Safety Authority and European Centre for Disease Prevention and Control. The European Union summary report on antimicrobial resistance in zoonotic and indicator bacteria from humans, animals and food in 2016. EFSA J. 2018; https://doi.org/10.2903/j.efsa.2018.5182.
7. Ward MJ, et al. Time-scaled evolutionary analysis of the transmission and antibiotic resistance dynamics of *Staphylococcus aureus* clonal complex 398. Appl Environ Microbiol. 2014;80:7275–82.
8. De Vries SPW, et al. Phylogenetic analysis and antimicrobial resistance profiles of *Campylobacte*r spp. from diarrheal patients and chickens in Botswana. PLoS One. 2018;13:e0194481.
9. Liu YY, et al. Emergence of plasmid –mediated colistin resistance mechanisms MCR-1 in animals and human beings in China: a microbiological and molecular biological study. Lancet Infecty Dis. 2016;16:161–8.
10. Tang KL, Caffrey NP, Nobrega DB, et al. Restricting the use of antibiotics in food-producing animals and its association with antibiotic resistance in food-producing animals and human beings: a systematic review and meta-analysis. Lancet Planet Health. 2017;1:e316–27.
11. Van Bunnik B, Woolhouse M. Modeling the impact of curtailing antibiotic usage in food animals on antibiotic resistance in humans. R Soc Open Sci. 2017;4:161067.
12. Forsman M, Haggerstrom B, Lindgren L, Jaurin B. Molecular analysis of β-lactamases from four species of *Streptomyces:* comparison of amino acid sequences with those of other β-lactamases. Microbiology. 1990;136:589–98.
13. Ogawara H, Kawamura N, Kudo T, Suzuki K-L, Nakase T. Distribution of β-lactamases in actinomycetes. Antimicrob Agents Chemother. 1999;43:3014–7.
14. Livermore D, Canton R, Gniadkowski M, et al. CXT-M changing the face of ESBLs in Europe. J Antimicrob Chemother. 2007;59:165–74.
15. Davies J, Davies D. Origins and evolution of antibiotic resistance. Microbiol Molecular Biol Rev. 2010;74:417–33.
16. Liu B, Pop M. ARDB—antibiotic resistance genes database. Nucleic Acid Res. 2009;37:D443–7.
17. Forsberg KJ, Patel S, Gibson MK, Lauber CL, Knight R, Fierer N, Daniels G. Bacterial phylogeny structure soil r5esistomes across habitats. Nature. 2014;509:612–6.
18. Etienne R, Ghozlane A, Tap J, et al. Prediction of the intestinal resistome by a three-dimensional structure-based method. Nat Microbiol. 2018; https://doi.org/10.1038/s41564-018-0292-6.
19. Willems RPJ, van Dijk K, Ket JCF, Vanderdroucke-Grauls CMJE. Evaluation of the association between gastric acid suppression and risk of intestinal colonization with multi-resistant microorganisms. A systematic review and meta-analysis. JAMA Intern Med. 2020; https://doi.org/10.1001/jamaiternmed.2020.0009.
20. Kapoor G, Saigal S, Elongavan A. Action and resistance mechanisms of antibiotics: a guide for clinicians. J Anaesthesiol Clin Pharmacol. 2017;33:300–5.
21. Eyal Z, Matzov D, Krupkin M, et al. A novel pleuromutilin antibacterial compound, its binding mode and selectivity mechanisms. Sci Rep. 2016;6:1–8.

22. Zaman SB, Hussain MA, Nye R, Mehta V, Mamun KT, Hossain N. A review on antibiotic resistance: alarm bells are ringing. Cureus. 2017;9:e1403.
23. Lambert PA. Cellular impermeability and uptake of biocides and antibiotics in gram-positive bacteria and mycobacteria. J Appl Microbiol. 2002;92(Suppl):46S. 54S.
24. Tenover FC. Mechanisms of antimicrobial resistance in bacteria. Am J Med. 2006;119(6 Suppl 1):S3–10.
25. Grundmann H, Aires-de-Sousa M, Boyce J, Tiemersma E. Emergence and resurgence of methicillin-resistant *Staphylococcus aureus* as a public health threat. Lancet. 2006;368:874–85.
26. Giedraitiene A, Vikauskiene A, Naginiene R, Pavilonis A. Antibiotic resistance mechanisms of clinically important bacteria. Medicina (Kaunas). 2011;47:137–46.
27. Higgins PG, Fluit AC, Schmitz FJ. Fluoroquinolones: structure and target sites. Curr Drug Targets. 2003;4:181–90.
28. Peterson E, Kaur P. Antibiotic resistance mechanisms in bacteria: relationship between resistance determinants of antibiotic producers, environmental bacteria, and clinical pathogens. Front Microbiol. 2018; https://doi.org/10.3389/fmicb.2018.02928.
29. Mojica MF, Bonomo RA, Fasi W. B1-metallo-beta lactamases: where do we stand? Curr Drug Targets. 2016;17:1029–50.
30. Perez F, Endimiani A, Hujer KM, Bonomo RA. The continuing challenge of ESBLs. Curr Opin Pharmacol. 2007;7:459–69.
31. Wright GD. Q & A: antibiotic resistance: where does it come from and what can we do about it? BMC Biol. 2010;8:123.
32. McMurry L, Petrucci RE Jr, Levy SB. Active efflux of tetracycline encoded by four genetically different tetracycline resistance determinants *in Escherichia coli*. Proc Natl Acad Sci USA. 1980;77:3974–7.
33. Nikaido H, Takatsuka Y. Mechanisms of RND multidrug efflux pumps. Biochem Biophys Acta. 2009;1794:769–81.
34. Van Bambeke F, Michot JM, Tulkens PM. Antibiotic efflux pumps in eukaryotic cells: occurrence and impact on antibiotic cellular pharmacokinetics, pharmacodynamics and toxicodynamics. J Antimicrob Chemother. 2003;51:1067–77.
35. Blanco P, Hernando-Amado S, Reales-Calderon J, et al. Bacterial multidrug efflux pumps: much more than antibiotic resistance determinants. Microorganisms. 2016;4:14. https://doi.org/10.3390/microorganisms4010014.
36. Fernandez L, Breidenstein EB, Hancock RE. Creping baselines and adaptive resistance to antibiotics. Drug Resis Updates. 2011;14:1–21.
37. Anderson D, Hughes D. Selection and transmission of antibiotic resistant bacteria. Microbiol Spectr. 2017;5:1–17.
38. D'Costa VM, McGrann KM, Hughes DW, Wright GD. Sampling the antibiotic resistome. Science. 2006;311:374–7.
39. Dantas G, Summer MOA, Oluwasegun RD, Church GM. Bacteria subsisting on antibiotic. Science. 2008;320:100–3.
40. Woodford N, Ellington MJ. The emergence of antibiotic resistant by mutation. Clin Microbiol Infect. 2007;13:5–18.
41. Schrag SJ, Perrot V, Levin BR. Adaptation to the fitness costs of antibiotic resistance in *Escherichia coli*. Proc R Soc Lond B Biol Sci. 1997;264:1287–91.
42. Poole K. Multidrug efflux pumps and antimicrobial resistance in *Pseudomonas aeruginosa* and related organisms. J Med Microbiol Biotechnol. 2001;3:255–64.
43. Heritage J, M'Zali FH, Gascoyne-Binzi D, HawKey PM. Evolution and spread of SHV extended-spectrum β-lactamases in gram-negative bacteria. J Antimicrob Chemother. 1999;44:309–18.
44. Huddleston JR. Horizontal gene transfer in the human gastrointestinal tract: potential spread of antibiotic genes. Infect Drug Resis. 2014;7:167–76.
45. Arcilla MS, van Hattem JM, Matamoros S, et al. Dissemination of themer-1colistin resistance gene. Lancet Infect Dis. 2015;16:147–9.

46. von Wintersdorff CJH, Penders J, van Niekerk JM, et al. Dissemination of antimicrobial resistance in microbial ecosystem through horizontal gene transfer. Front Microbiol. 2016; https://doi.org/10.3389/fmicb.2016.00173.
47. Domingues S, Harms K, Fricke WF, Johnsen PJ, da Silva GJ, Nielsen KM. Natural transformation facilitates transfer of transposons, integrons and gene cassettes between bacterial species. PLoS Pathog. 2012;8:e1002837. https://doi.org/10.1371/journal.ppat.1002837.
48. Friedman ND, Temkin E, Carmeli Y. The negative impact of antibiotic resistance. Clin Microbiol Infect. 2016;22:416–22.
49. O'Neill J. Tackling drug-resistant infections globally: final report and recommendations. Review Antimicrob. Resistance 2016; https://amr-review.org/. Accessed 17 Apr 2020.
50. Centers for Disease Control and Prevention. Antibiotic resistance threats in the United States. 2013; https://www.cdc.gov/drugresistance/threat-report-2013/pdf/ar-threats-2013-508-pdf
51. Jernigan JA, Kelly M, Hatfield MSPH, et al. Multidrug–resistant bacterial infections in US hospitalized patients, 2012–2017. N Engl J Med. 2020;382:1309–19.
52. Antony HA, Parija SC. Antimalarial drug resistance: an overview. Trop Parasitol. 2016;6:30–41.
53. Prestinaci F, Pezzotti P, Pantosti A. Antimicrobial resistance: a global multifaceted phenomenon. Pathog Glob Health. 2015;109:309.
54. Chokshi A, Sifri Z, Cennimo D, Horng H. Global contributors to antibiotic resistance. J Glob Infect Dis. 2019;11:36–42.
55. Shrestha P, Cooper BS, Coast J, et al. Enumerating the economic cost of antimicrobial resistance per antibiotic consumed to inform the evaluation of interventions affecting their use. Antimicrob Resist Infect Control. 2018;7:98. https://doi.org/10.1186/s13756-018-0384-3.
56. Thorpe KE, Joski P, Johnston KJ. Antibiotic-resistant infection treatment costs have doubled since 2002, now exceeding $2 billion annually. Health Aff. 2018;37:662–9.
57. Asiam B, Wang W, Arshad MI, et al. Antibiotic resistance: a rundown of a global crisis. Infect Drug Resist. 2018;11:1645–58.
58. Drug-resistant infections a threat to our economic future; 2017. Available from: www.worldbank.org. Accessed 21 Apr 2020.
59. Antibiotic resistance threats in the United States, 2013. https://www.cdc.gov/drugresistance/pdf/ar-threats-2013-508.pdf. Accessed 21 Apr 2020.
60. Anderson M, Clift C, Schulze K, et al. Health systems and policy analysis- averting the AMR crisis: what avenues for policy. Eur Obs Heal Sys Policies. 2019; [Google Scholar]
61. Utt E, Wells C. The global response to the threat of antimicrobial resistance and the important role of vaccines. Pharm Policy Law. 2016;18:179–97. https://doi.org/10.3233/PPL-160442.
62. Dadgostar P. Antimicrobial resistance: implications and costs. Infect Drug Resist. 2019;12:3903–10.
63. Animal production I antimicrobial resistance I food and agriculture organization of the United Nations. http://www.fao.org/antimicrobial-resistance/key-sectors/animal-production/en/
64. Buchy P, Ascioglu S, Buisson Y, et al. Impact of vaccines on antimicrobial resistance. Int J Infect Dis. 2020;90:188–96.
65. Hicks LA, Bartoces MG, Roberts RM, et al. US outpatient antibiotic prescribing variation according to geography, patient population, and provider speciality in 2011. Clin Infect Dis. 2015;60:1308–16.
66. Linder JA, Meeker D, Fox CR, Friedberg MW, Persell SD, Goldstein NJ, Doctor JN. Effects of behavioral interventions on inappropriate antibiotic prescribing in primary care 12 months after stopping the interventions. JAMA. 2017;318:1391.
67. Daneman N, Gruneir A, Newman A, et al. Antibiotic use in long-term care facilities. J Antimicrob Chemother. 2011;66:2856–63.
68. Rhee SM, Stone ND. Antimicrobial stewardship in long-term care facilities. Infect Dis Clin N Am. 2014;28:237–46.
69. US Food and Drug Administration. Battle of the bugs: fighting the antibiotic resistance. Updated May 4, 2016.; https://www.fda.gov//drugs/resourcesforyou/consumers/ucm43568.htm

70. Briceland LL, Nightingale CH, Quintiliani R, et al. Antibiotic streamlining from combination therapy to monotherapy utilizing an interdisciplinary approach. Arch Intern Med. 1988;148:2019–22.
71. Owens RC. Antimicrobial stewardship: concepts and strategies in the 21st century. Diag Microbiol Infect Dis. 2008;61:110–28.
72. Davey P, Marwick CA, Scott CL, et al. Interventions to improve antibiotic practices for hospital inpatients. Cochrane Database Syst Rev. 2017;2:CD003543. https://doi.org/10.1002/14651858.CD003543.pub4.
73. Nathwani D, Varghese D, Stephens J, Ansari W, Martin S, Charbonneau C. Value of hospital antimicrobial stewardship programs [ASPs]: a systemic review. Antimicrob Resist Infect Control. 2019;8:35. https://doi.org/10.1186/s13756-019-0471-0.
74. Vaughn VM, Flanders SA, Snyder A, et al. Excess antibiotic treatment duration and adverse events in patients hospitalized with pneumonia. Ann Intern Med. 2019;171:153–63.
75. Manyi-Loh C, Mamphwell S, Meyer E, Okoh A. Antibiotic use in agriculture and its consequential resistance in environmental sources: potential public health implications. Molecules. 2018;23:795. https://doi.org/10.3390/molecules23040795.

# Part I
# New Antibacterial Agents

# Chapter 2
# New Cephalosporins: Fifth and Sixth Generations

## 2.1 Introduction

The cephalosporins are synthetic antibiotics that were introduced in the 1970s and widely used to treat gram-positive and gram-negative bacteria infections globally due to their safety profile and bactericidal activity. They were introduced incrementally in stages with increasing spectrum of activity and are designated as first to fifth generations. The first-generation cephalosporins have good activity against gram-positive cocci but limited activity against gram-negative pathogens (i.e., cefazolin and cephalexin). Second-generation cephalosporins have improved gram-negative activity (e.g., cefotetan) and anaerobic coverage (e.g., cefoxitin). The third-generation cephalosporins have further improved gram-negative spectrum but decreased or variable gram-positive activity (e.g., cefotaxime, ceftriaxone, and ceftazidime). Cefepime and cefpirome are fourth-generation agents with good activity against most important gram-positive cocci, except methicillin-resistant *Staphylococcus aureus* and enterococci, and gram-negative bacilli.

## 2.2 Ceftaroline

Ceftaroline, a fifth-generation novel cephalosporin, is the first β-lactam agent shown to have good in vitro and in vivo activity against methicillin-resistant *Staphylococcus aureus* (MRSA) and was approved by the Food and Drug Administration (FDA) in the United States (US) in 2010 and by the European Commission in 2012 for the treatment of skin and soft tissue infection, including those caused by MRSA. It is also approved for acute bacterial community-acquired pneumonia (CAP).



I. W. Fong, *New Antimicrobials: For the Present and the Future*, Emerging Infectious Diseases of the 21st Century,
https://doi.org/10.1007/978-3-031-26078-0_2

### 2.2.1 Mechanism of Action and Microbial Spectrum

Similar to other cephalosporins and penicillins, ceftaroline bactericidal activity is mediated by binding to penicillin-binding proteins (PBPs), membrane-bound enzymes, resulting in impaired cell wall synthesis leading to cell lysis and death. Most β-lactam agents have low activity against MRSA, due to low affinity for binding to PBP2a (encoded by *mecA* gene), whereas ceftaroline has high affinity for binding PBP2a [1]. Against methicillin-sensitive *S. aureus* (MSSA), ceftaroline minimum inhibitory concentration for 90% ($MIC_{90}$) of isolates is 0.25 μg/ml and for MRSA the $MIC_{90}$ is 1–2 μg/ml [2]. It also has high activity against streptococci, including penicillin-resistant *S. pneumoniae* (PRSP) or multidrug-resistant strains (MDRSP), defined as penicillin resistant (MIC ≥8 μg/ml) and resistance to one or more other antibiotics (macrolide, tetracycline, levofloxacin, and trimethoprim/sulfamethoxazole [TMP/SMX]) [3]. Ceftaroline is 16-fold more active than ceftriaxone, with 98.7% of *S. pneumoniae* strains inhibited by 0.25 μg/ml [3]. It also has high binding to PBP 2b, 2x, and 1a which are responsible for PRSP. Ceftaroline has comparable activity to third-generation cephalosporins against aerobic gram-negative bacteria, but it is inactive against extended-spectrum β-lactamase *Enterobacteriaceae* (ESBL) and gram-negative anaerobes, such as *Bacteroides fragilis* and *Prevotella* spp. [4].

### 2.2.2 Pharmacokinetics of Ceftaroline

Ceftaroline fosamil is a prodrug which is rapidly dephosphorylated into the active ceftaroline after intravenous (IV) administration, and it has a linear pharmacokinetics profile. The plasma protein binding of the drug is about 20% and the terminal elimination half-life (t1/2) approximately 2.5 h, primarily eliminated by renal excretion, and the steady-state volume of distribution is 20.3 L [1]. The recommended dose of ceftaroline in adults is 600 mg IV over an hour every 12 h with a maximum concentration ($C_{max}$) of 21 μg/ml and the area of the concentration-time curve of 56 μg h/ml [4]. Dosage adjustment is needed for moderate to severe renal dysfunction; for moderate renal impairment (creatinine clearance [CrCl] >30 to ≤50 ml/min), the dose should be 400 mg every 12 h and for severe renal impairment (CrCl ≥15 to ≤30 ml/min) 300 mg every 12 h; and for those on hemodialysis 200 mg every 12 h after dialysis [5]. About 88% of the dose is recovered in the urine and 6% in the feces in 48 h. Ceftaroline penetration into the cerebrospinal fluid (CSF) was estimated to be 15% in inflamed meninges and 3% in the noninflamed meninges [4].

### 2.2.3 Clinical Efficacy of Ceftaroline

Multinational, double-blind, randomized studies in skin and soft tissue infections had shown comparable efficacy of ceftaroline compared to vancomycin and aztreonam [4]. These included subjects with *S. aureus* bacteremia (MSSA and MRSA). Thus, ceftaroline is a safe and effective option for severe skin/soft tissue infection requiring parenteral therapy for MRSA and gram-negative enteric infections. It would be interesting to determine if it were more effective than the inexpensive fixed combination, trimethoprim/sulfamethoxazole [TMP/SMX] which can be given parenterally and orally.

Ceftaroline has been compared to ceftriaxone for community-acquired pneumonia (CAP) in five randomized controlled trials (RCTs). A recent systematic review and meta-analysis included 1153 and 1050 control patients. The clinical efficacy and adverse effects were similar for the two groups [6]. However, the study concluded there was no significant difference for each of the following pathogens: *S. pneumoniae*, *S. aureus*, *H. influenza*, *H. parainfluenzae*, *Escherichia coli*, and *Klebsiella pneumoniae*, the number of isolates were not provided and commonly in CAP <30% of cases have an identified pathogen. Moreover, there were insufficient numbers of *S. aureus* pneumonia, particularly MRSA to determine its value for this subtype. Despite the lack of prospective studies of the value of ceftaroline for MRSA pneumonia, it has been argued that this agent is a potentially useful treatment option for MRSA CAP and associated bacteremia [7]. In another review of ceftaroline efficacy in pneumonia not restricted to CAP or RCT, but included observation studies, hospital-acquired pneumonia (HAP), and ventilator-associated pneumonia (VAP), the pooled efficacy was reported as 81.2%, and for MSSA and MRSA infections, it was >70% [8].

There is lack of prospective RCT of ceftaroline for severe MRSA infections, but a recent review of published case series and retrospective studies was recently published. This review included 22 papers published between 2010 and 2016 with 379 patients treated with ceftaroline for severe MRSA infections [9]. Although the overall cure rate was 74%, patients with MRSA bacteremia associated with infective endocarditis or pneumonia had high clinical failure rates of 30.3% and 27.6%, respectively. MRSA HAP and VAP treated for 7–8 days had cure rates of 58.3% and 57.1%, respectively. Only three patients had CNS MRSA infection: bacteremia with meningitis, ventricular-peritoneal shunt infection, and epidural abscess, all were cured with prolonged courses but one case received combination with rifampin.

### 2.2.4 Side Effects of Ceftaroline

The side effects of ceftaroline are similar to other cephalosporins such as ceftriaxone. Potential allergic reactions have occurred in 5.8% with rashes of 1.1–3.2%, generalized pruritus 2.2–3.5%, eosinophilia 1.6%, diarrhea 4.2–4–9%, *Clostridium difficile* colitis 1.3%, nausea 5.9%, asymptomatic positive Coombs test 10.7%, renal toxicity ≤2%, hepatic toxicity ≤3%, and neutropenia [4, 9]. Agranulocytosis of up to 13% has been reported with prolonged therapy greater than 7 days [9]. Ceftaroline, similar to other β-lactam agents, has been reported to cause encephalopathy in the presence of renal impairment. Among 28 patients with renal dysfunction (CrCl ≤30 ml/min), 3 developed encephalopathy associated with higher doses than recommended for severe infection [10].

The effect of ceftaroline on the normal bowel flora has been assessed in 12 healthy subjects and no measurable concentrations in the feces were found over 21 days [11]. There was minor alteration of the number of *E. coli* strains, moderate decreases in numbers of bifidobacteria and lactobacilli during the first 7 days, and increased numbers of clostridia during the same period, but no impact on bacteroides was found. Although no major impact was found on the intestinal flora, it would be of interest to determine its effect on the normal flora of the mouth, nares, skin, and vagina. Moreover, culture methods are inadequate to assess alterations of the normal microbiome as most of the microbes in the bowel are noncultureable.

## 2.3 Ceftobiprole

Ceftobiprole is another 5th-generation cephalosporin with broad spectrum against gram-negative and gram-positive bacteria, including MRSA, very similar to ceftaroline but superior pseudomonas in vitro activity. It is approved for CAP and HAP, excluding VAP, in several European and non-European countries but not the US. Basilea Pharmaceutica is expected to launch ceftobiprole in the US by 2022 (PMLIVE).

### 2.3.1 *Mechanism of Action and Microbial Activity of Ceftobiprole*

Like other β-lactam agents, ceftobiprole induce its bactericidal activity by inhibition of the transpeptidase moiety of the PBPs, with good binding to PBP2a of MRSA and PPB2x of PRSP [12]. Its activity against *S. aureus* (MSSA and MRSA) and *S. pneumoniae* (including MRSP) are very similar to ceftaroline, but it is more active against *Enterococcus faecalis*. In a recent study of large numbers of clinical isolates from hospitalized patients in the US from 2016, *S. aureus* isolates (including 1260 MRSA) were 99.3% susceptible, coagulase staphylococci ($n$ = 703) were 100% susceptible, *S. pneumoniae* ($n$ = 698) were 99.7% susceptible, and *E. faecalis* ($n$ = 347) were 100% susceptible [13]. Similar to ceftaroline, the in vitro activity of ceftobiprole against *Enterobacteriaceae* is very good, *E. coli* (99.8% susceptible), and *K. pneumoniae* (99.8% susceptible), but not against ESBL phenotypes. Among

**Table 2.1** In vitro susceptibility of ceftaroline and ceftobiprole for gram-positive bacteria

| Organisms and agents | $MIC_{90}$ [μg/ml] | % Susceptible |
|---|---|---|
| *S. aureus (MSSA)* | | |
| Ceftaroline | 0.25–1 | 100 |
| Ceftobiprole | 0.5–2 | 100 |
| Cefazolin | 1.0 | 100 |
| TMP-SMX | ≤0.12 | 97.9–99.7 |
| Vancomycin | 1.0 | 100 |
| *MRSA* | | |
| Ceftaroline | 1.0 | 96.4–100 |
| Ceftobiprole | 2.0 | 99.3–100 |
| TMP-SMX | ≤0.12–0.5 | 95.8 |
| Vancomycin | 1.0 | 100 |
| *MRCNS* | | |
| Ceftaroline | 0.5 | 100 |
| Ceftobiprole | 1–2 | 100 |
| TMP-SMX | 4–8 | 26.3–56.8 |
| Vancomycin | 2.0 | 100 |
| *S. pneumoniae* | | |
| Ceftaroline | 0.03–0.12 | 100 |
| Ceftobiprole | 0.30–0.5 | 100 |
| Ceftriaxone | 0.12–1 | 86.2–99 |
| Penicillin | 0.12–2 | 64.2–88 |
| *E. faecalis* | | |
| Ceftobiprole | 2.0 | 100 |
| Ceftaroline | 8.0 | NA |
| Ampicillin | 2.0 | 100 |
| Vancomycin | 2.0 | 96.3–100 |

Note: Data obtained from reference no. 13 and Karlowsky et al. Antimicrob Agents Chemother 2011; 55: 2837–40
Abbreviations: *MRCNS* methicillin-resistant coagulase-negative staphylococci, *TMP-SMX* trimethoprim-sulfamethoxazole

*P. aeruginosa* isolates ($n = 1017$), 72.7% were susceptible to ceftobiprole compared to 86% susceptible to ceftazidime. Table 2.1 compares the susceptibility of ceftaroline and ceftobiprole.

### 2.3.2 Pharmacology of Ceftobiprole

The water-soluble ceftobiprole medocaril (prodrug) is rapidly activated in plasma by type A esterase to the active moiety [14]. Ceftobiprole protein binding is only 16% and the volume of distribution is similar to the extracellular fluid compartment;

the mean half-life is 3.1 h, and about 83% of the drug is excreted in the urine [14]. The recommended dosage is 500 mg every 8 h infused over 2 h, and dosage adjustment is needed for moderate to severe renal impairment: 500 mg every 12 h for CrCl 30–50 ml/min, 250 mg every 12 h for CrCl <30 ml/min, and 250 mg every 24 h for CrCl <30 ml/min or end-stage renal failure on intermittent hemodialysis [15].

As with other β-lactam agents, ceftobiprole exhibits time-dependent bacterial activity and in experimental models plasma concentrations above the MIC for 30–60% of the dosing interval [t > MIC] resulted in effective bacterial killing of >2–4 $\log_{10}$ of *S. aureus*, *S. pneumoniae*, and *Enterobacteriaceae* [14]. Data from RCT found that 51% of t > MIC correlated with favorable outcomes [16].

### 2.3.3 Phase 3 Clinical Studies of Ceftobiprole

Two RCT phase 3 trials for the treatment of CAP and HAP were performed with ceftobiprole. The first double-blind, non-inferiority, RCT compared ceftobiprole 500 mg every 8 h (q8h) to ceftriaxone 2 g once daily (optional linezolid for suspected MRSA) for 7 days in CAP and found no difference in clinical cure (76.4% vs 79.3%), meeting the non-inferiority target [17]. In a RCT (non-inferiority) of HAP ($n$ = 571) and VAP ($n$ = 210), patients were given ceftobiprole or ceftazidime/linezolid for 7–14 days. The clinical efficacy of the two groups were similar in HAP, 77.8% and 76.2%, and met the non-inferiority criteria; but in VAP clinical cure was less with ceftobiprole by 13.7% (37.7% vs 55.9%) and did not meet the non-inferiority goal [18].

Ceftobiprole has been studied in two RCTs for complicated skin/soft tissue infections due to gram-positive bacteria, compared to vancomycin, and those due to gram-positive and gram-negative bacteria compared to vancomycin and ceftazidime with similar cure rates as the comparators [90–93.3%] [19, 20]. A third RCT of complicated skin/soft tissue infections is ongoing to compare ceftobiprole vs vancomycin and aztreonam (NCT03137173).

There is limited data on the efficacy of ceftobiprole in *S. aureus* (MRSA) bacteremia or infective endocarditis, osteomyelitis, and prosthetic joint infections to draw any conclusion [14].

### 2.3.4 Safety Profile of Ceftobiprole

Overall, the safety profile of ceftobiprole is similar to other cephalosporins with the most common adverse events being nausea, vomiting, diarrhea, infusion site reactions, dysgeusia, and hypersensitivity reactions being the most common [14]. *C. difficile* colitis is rare and in experimental models is less likely than other cephalosporins (cefoxitin, ceftriaxone, cefotaxime, ceftazidime) and carbapenems [21], probably due to absence of fecal excretion, minimal effect on the normal bowel flora, and inhibitory effect on *C. difficile* [22]. Unlike ceftaroline, neutropenia has not been a significant problem with prolonged therapy and neurotoxicity has not been reported [22].

### *2.3.5 Role of Ceftaroline and Ceftobiprole in Clinical Practice*

The clinical niche for ceftobiprole and ceftaroline in the management of infections is difficult to define and rationalize, despite their novel activity against MRSA. For severe CAP, their efficacy is no better than standard therapy with ceftriaxone and coverage for MRSA is not routinely recommended. Both agents may be suitable for empiric therapy of HAP (excluding VAP) but their lack of activity against ESBL is of concern, as well as inadequate data on severe MRSA infection with bacteremia. These agents could be suitable for mixed severe skin/soft tissue infections with MRSA, streptococci, and coliforms instead of combined therapy with vancomycin and may be cost-effective especially in the presence of renal impairment. Predictably clinical usage of ceftaroline (likely the same for ceftobiprole) has selected for resistance in MRSA with mutations leading to amino acid substitutions close to the active site and at the allosteric site of PBP2 that interfere with binding [23–25].

## 2.4 Cefiderocol: A Novel Siderophore Cephalosporin

Cefiderocol is the latest novel 5th–6th generation cephalosporin to be approved by the FDA in November 2019 for the treatment of complicated urinary tract infections (UTIs) and pyelonephritis, and in September 2020 it was approved for hospital-acquired and ventilator-associated bacterial pneumonia (HAP/VAP). This agent was designed to combat multiresistant gram-negative bacterial infections including the globally increasing carbapenem-resistant strains.

### *2.4.1 Chemistry and Mechanisms of Action of Cefiderocol*

Siderophores are natural iron-chelating molecules produced by bacteria to allow transport of iron into their cells for growth and survival. Combining siderophore with synthetic β-lactams to increase bacterial cell entry and, hence, killing have been in development since the 1980s [26]. The conjugate of iron-siderophore-antibiotic complex is actively transported by iron transporter outer membrane protein ("Trojan horse' strategy) into the periplasmic space of the bacterial cell, dissociates, and allows bacterial killing. Cefiderocol has a pyrrolidinium group on the C-3 side chain (similar to cefepime) and a carboxypropanoxyimino group on the

C-7 side chain (similar to ceftazidime), with increased potency against gram-negative bacteria and stability against β-lactamases and carbapenemases [26].

Similar to other β-lactam agents, cefiderocol inhibits cell wall synthesis by binding to PBPs with bactericidal effect. Due to its high binding affinity to PBP3, it has potent activity against *E. coli*, *K. pneumoniae*, *P. aeruginosa*, and *Acinetobacter baumannii* [27]. A pivotal aspect of the drug is the intrinsic structural stability to wide variety of serine- and metallo-β-lactamases including *K. pneumoniae* carbapenemase (KPC), oxacillin carbapenemase (OXA), New Delhi metallo-β-lactamase (NDM), and Verona integron-encoded metallo-β-lactamase (VIM) carbapenemase [26]. Moreover, the active transport siderophore mechanism has allowed cefiderocol to overwhelm drug resistance with permeability barrier due to porin channel loss or overexpression of multidrug efflux pumps.

## 2.4.2 In Vitro Activity of Cefiderocol

The in vitro activity of cefiderocol against key gram-negative bacteria associated with hospital-acquired infections (HAP, VAP), bacteremia, and complicated UTI (including difficult to treat bacterial species) has been assessed by large multinational surveillance studies (SIDERO-WT studies 2014, 2015, 2016) from North America and Europe [28]. Based on preclinical in vivo efficacy and pharmacokinetic-pharmacodynamic data, the provisional breakpoints approved by CLSI for susceptible, intermediate, and resistant were 4, 8, and 16 μg/ml, respectively, for *Enterobacteriaceae*, *P. aeruginosa*, *A. baumannii*, and *Stenotrophomonas maltophilia*. The SIDERO-WT program tested over 28,029 clinical isolates collected from ≈100 hospitals from 2014 to 2017. In each testing period, >99% of bacteria had low MIC values. The $MIC_{90}$ for *Enterobacteriaceae* ranged from 0.25 to 1 μg/ml, for the nonfermenters (*P. aeruginosa, Burkholderia cepacia, S. maltophilia*) the $MC_{90}$ ranged from 0.03 to 1 μg.ml, and 1–4 μg/ml for *A. baumannii* [28, 29]; see Table 2.2.

Cefiderocol has been tested against carbapenem-nonsusceptible and multidrug-resistant clinical isolates collected from 52 countries (in Europe, North and South America, Asia-Pacific, and Africa) from 2014 to 2016. Against 1022 strains of resistant *Enterobacteriaceae*, including 23.0% resistant to ceftazidime-avibactam and 22.2% colistin resistant, cefiderocol inhibited 97.0% at <4 μg/ml [29]. The $MIC_{90}$ for MDR *P. aeruginosa* ($n$ = 262) was 1 μg/ml, for *S. maltophilia* ($n$ = 217) was 0.25 μg/ml, and for MDR *A. baumannii* ($n$ = 368) was 8 μg/ml; with 99.2%, 100%, and 89.7% of the respective strains MICs ≤4 μg/ml of cefiderocol.

Since the approval of cefiderocol in 2019, the FDA has adopted alternative susceptibility breakpoints for *Enterobacteriaceae* and *P. aeruginosa*, ≤2 and ≤1 μg/ml, respectively, but interpretive criteria were not provided for *S. maltophilia* and *Acinetobacter* species [30]. This may create confusion for clinical laboratories as this is different from the CLSI guidelines. Moreover, MIC testing must be conducted in Mueller Hinton broth deleted of iron and cation adjusted.

**Table 2.2** In vitro susceptibility of cefiderocol and comparators against gram-negative bacteria

| Organisms and agents | $MIC_{90}$ [μg/ml] | % Susceptible |
|---|---|---|
| *Enterobacteriaceae* | | |
| Cefiderocol | 0.5–1.0 | >99 |
| Ceftobiprole | >16 | 82.5 |
| Ceftaroline | >32 | 75.5 |
| Ceftriaxone | >8 | 81.9 |
| Levofloxacin | >4 | 82.2 |
| Meropenem | 0.006 | 98.7 |
| Pip-Tazo | 16 | 91.6 |
| *P. aeruginosa* | | |
| Cefiderocol | 0.5 | 99.7–100 |
| Ceftobiprole | 16 | 72.7 |
| Ceftazidime | 16 | 86 |
| Imipenem | 8.0 | 78 |
| Pip-Tazo | 64 | 80.8 |
| Amikacin | 8.0 | 96.7 |
| *Carbapenem-nonsusceptible strains* | | |
| *Enterobacteriaceae* | | |
| Cefiderocol | 4.0 | 97 |
| Ceftazidime-avibactam | >64 | 77 |
| Ceftolozane-tazobactam | >64 | 1.7 |
| Ciprofloxacin | >64 | 11.5 |
| Colistin | >8.0 | 77.8 |
| *P. aeruginosa* | | |
| Cefiderocol | 1.0 | 99.2 |
| Ceftazidime-avibactam | >64 | 36.3 |
| Ceftolozane-tazobactam | >64 | 24.1 |
| Ciprofloxacin | >8.0 | 1.2 |
| Colistin | 1.0 | 99.6 |
| *A. baumannii* | | |
| Cefiderocol | 8.0 | 90.9 |
| Ceftazidime-avibactam | >64 | NA |
| Ceftolozane-tazobactam | >64 | NA |
| Ciprofloxacin | >8.0 | 0 |
| Colistin | 1.0 | 94.6 |
| *S. maltophilia* | | |
| Cefiderocol | 0.25 | 100 |
| Ceftazidime-avibactam | 64 | NA |
| Ceftolozane-tazobactam | 64 | NA |
| Ciprofloxacin | >8.0 | 0 |
| Colistin | >8.0 | NA |

### 2.4.3 Resistance to Cefiderocol

Ongoing surveillance program shows that a low proportion of gram-negative bacteria are nonsusceptble to cefiderocol with MIC >4 µg/ml, 0.4–0.7% from 2014 to 2017 [28]. The most frequent species are *A. baumannii,* mainly *Pseudomonas* extended resistant (PER) β-lactamase producers, and NDM-producing *Enterobacteriaceae*. Investigations revealed that cefiderocol resistance could be reversed by β-lactamase inhibitors (BLIs), although NDM or PER production alone might not be sufficient to cause cefiderocol resistance [31].

### 2.4.4 Pharmacokinetic/Pharmacodynamic Aspects of Cefiderocol

Cefiderocol has an elimination half-life of 2–3 h with 98.6% excreted in the urine and 2.8% in the feces and protein binding of 58% [32]. Single dose infusion of 2 g produces $C_{max}$ of 89.7 (over 3 h) to 156 µg/ml [1 h] and AUC 386–386 µg X h/ml. As with other β-lactam agents, it exhibits time-dependent bactericidal activity, and extended infusion of these agents may be more effective than bolus infusion in critically ill patients [33]. Based on experimental animal models with difficult to treat gram-negative bacteria and the Monte Carlo simulation model, 2 g cefiderocol infused over 3 h every 8 h was selected as a standard dose to achieve 75% T > MIC for strains of bacteria with MIC ≤4 µg/ml [32]. This is expected to produce bactericidal activity with ≥1 log reduction for 90% of the target bacteria.

Dose adjustment is needed for patients with moderate to severe renal impairment: 1.5 g every 8 h for CrCl 30 to <60 ml/min, 1 g every 8 h for CrCl 15 to <30 ml/min, 0.75 g every 12 h for CrCl <15 ml/min, 0.75 g every 12 h on intermittent hemodialysis with extra-dose after dialysis, and 1.5 g every 12 h for continuous hemodialysis [32].

### 2.4.5 Efficacy and Tolerability of Cefiderocol

Cefiderocol and other recently marketed antibiotics have been developed to counter the growing number of resistant gram-negative bacterial infections (especially carbapenem-resistant strains), including *Enterobacteriaceae*, *P. aeruginosa*, *A. baumannii*, and *S. maltophilia.* However, it is difficult to design a trial for carbapenem-resistant bacteria due to the relative sparsity. In a phase 2, multicenter, double-blind, non-inferiority RCT (APEKS-cUTI Study) at 67 hospitals in 15 countries, 448 patients with complicated UTI or acute uncomplicated pyelonephritis at risk for multidrug-resistant gram-negative infections were randomized to cefiderocol (2 g) or imipenem (1 g) every 8 h for 7–14 days [34]. Among 371 patients with qualifying gram-negative pathogens, 73% of 252 cefiderocol-treated and 55% of 119 imipenem-treated patients were cured (18.6% difference), $p = 0.0004$. Adverse events were primarily mild to moderate gastrointestinal (GI) side effects in 41% of cefiderocol recipients and 51% of imipenem recipients. There were fewer cases of *C. difficile* infection in the cefiderocol arm. However, prolonged treatment has been reported to result in neutropenia [30].

The CREDIBLE-CR study of critically ill patients with hospital-acquired multidrug-resistant (carbapenem-resistant) pneumonia, bacteremia, and sepsis randomized 150 patients (open-label), 101 to cefiderocol versus 49 to best available therapy [BAT] [35]. Cefiderocol had similar clinical and microbiological efficacy as BAT, but more deaths occurred in the cefiderocol group, mainly in those with *Acinetobacter* spp. infections. The deaths were not related to adverse events, but half the deaths were related to underlying comorbidity or infection complications other than the original gram-negative infection at randomization.

In a double-blind, non-inferiority RCT of patients with nosocomial pneumonia (HAP/VAP) conducted globally, cefiderocol ($n = 148$) was compared to meropenem ($n = 152$) (APEKS-NP study) [36]. The primary endpoint was the all-cause mortality at day 14. The baseline pathogens included *K. pneumoniae* (32%), *P. aeruginosa* (16%), *A. baumannii* (16%), and *E. coli* (14%). Sixty percent of patients were mechanically ventilated. The all-cause mortality was similar at 14 days, 28 days, and end of study, 12.4% with cefiderocol and 11.6% with meropenem. However, this study did not have sufficient carbapenem nonsusceptible or resistant infections.

### 2.4.6 Role of Cefiderocol in Clinical Infectious Diseases

The role of cefiderocol will be limited to treat infections with multiresistant gram-negative bacteria, particularly carbapenem-resistant strains which are still rare in North America but more prevalent elsewhere and are expanding globally. Its greatest use will be for therapy of complicated UTI and hospital-acquired sepsis including HAP/VAP with these resistant strains of bacteria, and this will increase our ability to provide safe and effective therapy. This agent could be used for sepsis and bacteremia with strains resistant to all other existing agents or only susceptible to colistin which is more toxic. Cefiderocol has recently been used as rescue therapy in 10 critically ill patients with VAP and bacteremia with carbapenem-resistant *A. baumannii, S. maltophilia*, and *K. pneumoniae* with 30-day clinical success of 70% and survival of 90% [37], but larger post-marketing data are needed on these types of cases.

However, it is surprising and concerning that cefiderocol had greater mortality than best available treatment in critically ill patients with bacteremia and severe sepsis with carbapenem nonsusceptible gram-negative infections. The meaning and significance of this is unclear, and further data and studies are needed. However, other clinical trials are in progress including comparison of cefiderocol compared to BAT for gram-negative bacteremia [30]. This drug should be of limited use for the mutiresistant gram-negative infections with oversight by antibiotic stewardship programs, as widespread empiric use will no doubt lead to increased cefiderocol-resistant species/strains. The recent Infectious Diseases Society of America (IDSA) guidelines for treatment CRE and MDR-*P. aeruginosa* list cefiderocol as an option for treatment of pyelonephritis and infection outside the urinary tract for infection due to these organisms [38].

The combined data from two recently completed phase 3 RCTs showed that cefiderocol was more effective than comparators of best available therapy for gram-negative bacteria producing metallo-β-lactamases [39]. In carbapenem-resistant *A. baumannii* bacteremia, the results have been mixed, one study showed cefiderocol produced lower mortality than colistin-containing regimens, and another study showed no significant lowering of mortality compared to colistin [40].

## References

1. Isikawa T, Matsunaga N, Tawada H, et al. TAK-599, a novel N-phosphono type prodrug of anti-MRSA cephalosporin T-91825: synthesis, physiochemical and pharmacological properties. Biorg. Med Chem. 2003;11:2427–37.
2. Welte T, Kantecki M, Stone GG, Hammond J. Ceftaroline fosamil as a potential treatment option for *Staphylococcus aureus* community-acquired pneumonia in adults. Int J Antimicrob Agents. 2019;54:410–22.
3. Pfaller MA, Farrell DJ, Sader HS, Jones RN. AWARE Ceftaroline surveillance program [2008–2010]: trends in resistance patterns among *Streptococcus pneumoniae, Haemophilus influenzae, and Moraxella catarrhalis* in the United States. Clin Infect Dis. 2012;55:S187–S93.
4. Shirley D-AT, Heil EL, Johnson JK. Ceftaroline fosamil: a brief clinical review. Infect Dis Ther. 2013;2:95–110.
5. TEFLARO$^R$ [ceftaroline fosamil], prescribing information. St. Louis: forest Pharmaceuticals, Inc., 2012.
6. Lan S-H, Chang S-P, Lai L-C, Chao C-M. Efficacy and safety of ceftaroline for the treatment of community-acquired pneumonia: a systematic review and meta-analysis of randomized controlled trials. J Clin Med. 2019;8:824. https://doi.org/10.3390/jcm8060824.
7. Welte T, Kantecki M, Stone GG, Hammond J. Ceftaroline fosamil as a potential treatment for *Staphylococcus aureus* community-acquired pneumonia in adults. Int J Antimicrob Agents. 2019;54:410–22.
8. Sotgiu G, Aliberti S, Gramegna A, et al. Efficacy and effectiveness of ceftaroline fosamil in patients with pneumonia: a systemic review and meta-analysis. Resp Res. 2018; https://respiratory-research.biomedcentral.com/articles/10.1186/s12931-018-0905-x.
9. Cosimi RA, Beik N, Kubliak DW, Johnson JA. Ceftaroline for severe methicillin-resistant *Staphylococcus aureus infections:* a systematic review. Open forum. Infect Dis. 2017;4:ofx084.
10. Martin TCS, Chow S, Johns ST, Mehta SR. Ceftaroline-associated encephalopathy in patients with severe renal impairment. Clin Infect Dis. 2020;70:2002–4.
11. Panagiotidis G, Bacstrom T, Asker-Hagelberg JA, Weintraub A, Nord CE. Effect of ceftaroline on normal human intestinal microflora. Animocrob Agents Chemother. 2010;54:1811–4.
12. Davies TA, Page MG, Shang W, et al. Binding of ceftobiprole and comparators to the penicillin binding proteins of *Escherichia coli, Pseudomonas aeruginosa, Staphylococcus aureus, and streptococcus pneumonia*. Antimicrob Agents Chemother. 2007;51:2621–4.

13. Pfaller MA, Flamm RK, Mendes RE, et al. Ceftobiprole activity against gram-positive and –negative pathogens collected from the United States in 2006 and 2016. Antimicrob Agents Chemother. 2019;63:e01566–18.
14. Giacobbe DR, Giuseppe de Rosa F, Del Bono V, et al. Ceftobiprole: drug evaluation and place in therapy. Expert Rev Anti-Infect Ther. 2019; https://doi.org/10.1080/1478210.2019.1667229.
15. Syed YYY. Ceftobiprole medocaril: a review of its use in patients with hospital- or community-acquired pneumonia. Drugs. 2014;74:1523–42.
16. Muller AE, Punt N, Mouton JW. Exposure to ceftobiprole is associated with microbiological eradication and clinical cure in patients with nosocomial pneumonia. Antimicrob Agents Chemother. 2014;58:2412–9.
17. Nicholson SC, Welte T, File TM Jr, et al. A randomized, double-blind trial comparing ceftobiprole medocaril with ceftriaxone with or without linezolid for the treatment of patients with community-acquired pneumonia requiring hospitalization. Int J Antimicrob Agents. 2012;39:240–6.
18. Awad SS, Rodriguez AH, Chuang YC, et al. A phase 3 randomized double-blind comparison of ceftobiprole medocaril versus ceftazidime plus linezolid for the treatment of hospitalized acquired pneumonia. Clin Infect Dis. 2014;59:51–61.
19. Noel GJ, Strauss RS, Amsler K, et al. Results of a double-blind, randomized trial of ceftobiprole treatment of complicated skin and skin structure infections caused by gram-positive bacteria. Antimicrob Agents Chemother. 2008;52:37–44.
20. Noel GJ, Bush K, Bagchi P, et al. A randomized, double-blind trial comparing ceftobiprole medocaril with vancomycin plus ceftazidime for the treatment of patients with complicated skin and skin-structure infections. Clin Infect Dis. 2008;46:647–55.
21. Nerandzic MM, Donsky CJ. Effect of ceftobiprole treatment on growth of and toxin production by *Clostridium difficile* in cecal contents of mice. Antimicrob Agents Chemother. 2011;55:2174–7.
22. Grau S. Safety and tolerability of ceftobiprole. Rev Esp Quimoter. 2019;32(Suppl 3):34–6.
23. Long SW, Olsen RJ, Mehta SC, et al. PBP2a mutation causing high-level ceftaroline resistance in clinical methicillin-resistant *Staphylococcus aureus* isolates. Antimicrob Agents Chemother. 2014;58:6668–74.
24. Lahiri SD, Alm RA. Potential of *Staphylococcus aureus* isolates carrying different PBP2a alleles to develop resistance to ceftaroline. J Antimicrob Chemother. 2016;71:34–40.
25. Shaumburg F, Peters G, Alabi A, et al. Misense mutations of PBP2a are associated with reduced susceptibility of ceftaroline and ceftobiprole in African MRSA. J Antimicrob Chemother. 2016;71:41–4.
26. Sato T, Yamawaki K. Cefiderocol: discovery, chemistry, and in vivo profiles of a novel siderophore cephalosporin. Clin Infect Dis. 2019;69: [Suppl 7]:S538–43.
27. Ito A, Sato T, Ota M, et al. In vitro antibacterial properties of cefiderocol, a novel siderophore cephalosporin, against gram-negative bacteria. Antimicrob Agents Chemother. 2017;62:e01454–17.
28. Yamano Y. In vitro activity of cefiderocol against a broad range of clinically important gram-negative bacteria. Clin Infect Dis. 2019;69(Suppl 7):S544–51.
29. Hackel MA, Tsuji M, Yamano Y, Echols R, Karlowsky JA, Sahm DF. In vitro activity of the siderophore cephalosporin, cefiderocol, against carbapenem-nonsusceptible and multidrug-resistant isolates of gram-negative bacilli collected worldwide in 2014 to 2016. Antimicrob Agents Chemother. 2018;62:e01968–17.
30. Shields RK. Case commentary: the need for cefiderocol is clear, but are the supporting clinic data? Antimirob Agents Chemother. 2020;64:e00059–20.
31. Ito A, Kuroiwa M, Ishioka Y, et al. Characterization of isolates showing high MICs to cefiderocol from the global surveillance study SIDERO-WT-2014. In: Poster presentation at: American Society of Microbiology annual meeting ASM-microbe. San Francisco, 20–24 June 2019, Poster AAR-774.

32. Katsube T, Eschols R, Wajima T. Pharmacokinetic and pharmacodynamics profiles of cefiderocol, a novel siderophore cephalosporin. Clin Infect Dis. 2019;69(Suppl 7):S552–8.
33. Vardakas KZ, Voulgaris GL, Maliaros A, Samonis G, Falagas ME. Prolonged versus short-term infusion of antipseudomonal β-lactams for patients with sepsis: a systematic review and meta-analysis of randomized trials. Lancet Infect Dis. 2018;18:108–20.
34. Portsmouth S, van Veenhuyzen D, Echols R, et al. Cefiderocol versus imipenem-cilastatin for the treatment of complicated urinary infections caused by gram-negative uropathogens: a phase 2, randomized, double-blind. Non-inferiority trial. Lancet Infect Dis. 2018;18:1319–28.
35. Bassetti M, Echols R, Matsunaga Y, et al. Efficacy and safety of cefiderocol or best available therapy for the treatment of serious infections caused by carbapenem-resistant gram-negative bacteria [CREDIBLE-CR]: a randomized, open-label, multicenter, pathogen-focused, descriptive, phase 3 trial. Lancet Infect Dis. 2021;21:226–40.
36. Wunderink RG, Matsunaka Y, Aryasu M, et al. Cefiderocol versus high-dose, extended-infusion meropenem for the treatment of gram-negative nosocomial pneumonia [APEKS-NP]: a randomized, double-blind, phase 3, non-inferiority trial. Lancet Infect Dis. 2021;21:213–25.
37. Falcone M, Tiseo G, Nicastro M, et al. Cefiderocol as rescue therapy for *Acinetobacter baumannii* and other carbapenem-resistant gram-negative infections in intensive care unit patients. Clin Infect Dis. 2021;72:2021–4.
38. Tammma PD, Aitken SL, Bonomo RA, Mathers AJ, van Duin D, Clancy JC. Infectious Diseases Society of America 2022 guidance on the treatment of extended-spectrum β-lactamase producing enterobacterales [ESBL-E], carbapenem-resistant Enterobacterales [CRE], and *Pseudomonas aeruginosa* with difficult-t—treat resistance [DTR-*P. aeruginosa*]. Clin Infect Dis. 2022;75:187–212.
39. Timsit J-F, Paul M, Shields RK, Echols R, Baba T, Yamano Y, Portsmouth S. Cefiderocol for treatment of infections due to metallo-β-lactamase-producing pathogens in the CREDIBLE-CR and APEKS-NP phase 3 randomized studies. Clin Infect Dis. 2022;75:1081–4.
40. Falcone M, Tiseo G. Cefiderocol for the treatment of metallo-β-lactamases producing gram-negative bacilli: light and shadows from the literature. Clin Infect Dis. 2022;75:1085–7.

# Chapter 3
# New β-Lactam-β-Lactamase Inhibitor Combinations

## 3.1 Introduction

Widespread use of β-lactam agents (penicillins, cephalosporins, carbapenems, and less frequently monobactams) have led to increasing global resistance, especially among gram-negative bacteria, largely due to β-lactamases. Two strategies have been used to overcome this dilemma: (1) design of novel β-lactam agents that are able to evade inactivation by β-lactamases and (2) combining β-lactamase inhibitors with the β-lactam agents. There are over 850 β-lactamases described which are divided in 4 distinct classes; A, C, and D are serine enzymes and class B comprises a heterogeneous group of zinc metalloenzymes (discussed in Chap. 1).

The three commonly used β-lactamase inhibitors are clavulanic acid (isolated from *Streptomyces clavuligerus* in 1970 and in use for >40 years), sulbactam, and tazobactam. These compounds have structural similarity with penicillin and are effective against many bacteria expressing class A β-lactamases (TEM, SHIV, and CTX-M), but not AmpC or KPC (*Klebsiella pneumoniae* carbapenemase) enzymes [1]. These penicillin-β-lactamase inhibitor combinations include amoxicillin-clavulanate (used mostly in the community), ampicillin-sulbactam (used for severe community-acquired infections admitted to hospital), and piperacillin-tazobactam (PIP/TAZO), used mainly for healthcare-associated infections including *Pseudomonas aeruginosa*. However, many gram-negative pathogens that produce extended-spectrum β-lactamases (ESBLs), carbapenemases, or multiple β-lactamases have become resistant to these combinations. Thus, newer classes of β-lactamase inhibitors with wider spectrum have been developed to combine with third- and later-generation cephalosporins and carbapenems to combat these more resistant bacteria.



I. W. Fong, *New Antimicrobials: For the Present and the Future*, Emerging Infectious Diseases of the 21st Century,
https://doi.org/10.1007/978-3-031-26078-0_3

## 3.2 Ceftolozane-Tazobactam

Ceftolozane is a new cephalosporin, structurally similar to ceftazidime, combined with the established β-lactamase inhibitor, tazobactam, to extend and protect its broad spectrum of activity. Ceftolozane is an oxyimino-cephalosporin (similar to ceftazidime) with a pyrazole (hydrothiazine) ring side chain at position 3 that increases stability against AmpC β-lactamases and prevent hydrolysis by *P. aeruginosa* [2]. The modification results in higher affinity and wider inhibition of the main penicillin-binding proteins (PBPs) of *P. aeruginosa* (PBP1b, PBP1c, PBP2, and PBP3) compared to ceftazidime, but lower affinity to PBP4 does not allow overexpression of AmpC [3]. Ceftolozane/tazobactam (Zerbaxa™) was approved in the United States (US) and European Union (EU) for complicated intra-abdominal infections with metronidazole and complicated urinary tract infections since 2014.

### 3.2.1 *Antimicrobial Activity of Ceftolozane/Tazobactam*

Ceftolozane is more stable to AmpC β-lactamase of *P. aeruginosa*, and a poor substrate for the Mex efflux pumps found in this species; thus it is the most potent anti-pseudomonas β-lactam agent. However, it is not stable to extended β-lactamases (ESBL), and combination with tazobactam extends its activity against some ESBL-producing *Enterobacteriaceae* and some *Bacteroides* spp. Tazobactam binds irreversibly to β-lactamases with slow hydrolysis and inhibits class A β-lactamases and the class C cephalosporinases. Ceftolozane/tazobactam (C/T) is stable to narrow spectrum β-lactamases (TEM-1, TEM-2, SHC-1, and OXA-1), but ESBLs (TEM-3-9, SHV-2-4, OXA-2, and CTX-M-3-18) reduces the activity, but it may still remain effective [2]. It has no activity against serine carbapenemases (KPC) or metallo-β-lactamase-producing organisms.

C/T has activity against most multidrug-resistant (MDR) *P. aeruginosa* and higher in vitro activity against *Enterobacteriaceae* than ceftazidime, cefepime, and piperacillin/tazobactam, but little or no activity against *Stenotrophomonas maltophilia* and *Acinetobacter* species [4]. It has good activity against β-hemolytic streptococci (*Streptococcus pyogenes* and *Streptococcus agalactiae*), modest activity against *Streptococcus pneumoniae*, no activity against staphylococci and enterococci, and activity against some anaerobes [4]. Table 3.1 shows the $MIC_{90}$ for different bacterial species and the breakpoints for susceptible, intermediate, and resistance.

**Table 3.1** In vitro activity of ceftolozane/tazobactam and breakpoints

| Organisms | $MIC_{90}$ µg/ml | Susceptible | Intermediate | Resistance [MIC, µg/ml] |
|---|---|---|---|---|
| *P aeruginosa* | 1.0 | <8 | 8 | ≥16 |
| *Enterobacteriaceae* | | ≤2 | 4 | ≥8 |
| *E. coli* | 0.5 | | | |
| *K. pneumoniae* | 8 | | | |
| *K. oxytoca* | 0.5 | | | |
| *E. cloacae* | 8 | | | |
| *Citrobacter* spp. | 8 | | | |
| *Proteus* spp. | 0.5–1 | | | |
| *Serratia* spp. | 1 | | | |
| *Streptococci* | | ≤8 | 16 | ≥32 |
| *S. pyogenes* | ≤0.12 | | | |
| *S. pneumoniae* | 8 | | | |
| *S. agalactiae* | 0.5 | | | |
| *Anaerobes* | | | | |
| *B. fragilis* | 2 | ≤8 | 16 | ≥32 |
| *Clostridium* spp. | 0.5 | NA | NA | NA |

Data obtained from reference [4]

### 3.2.2 Pharmacokinetics/Pharmacodynamics of Ceftolozane/Tazobactam

C/T mean elimination half-life is 2–3 h, cleared unchanged by the kidneys (92%), serum protein binding about 20%, and the volume of distribution is equivalent to the extracellular compartment [5]. The recommended dose in patients with normal renal function is 1000/500 mg every 8 h (q8h) infused over 1 h intravenously (IV). Dosage adjustment is needed for renal impairment with creatinine clearance (CrCl) below 50 ml/min as follows: CrCl 30–50 ml/min, 500 mg/250 mg q8h; CrCl 15–29 ml/min, 250/125 mg q8h; end-stage renal failure on hemodialysis, 500/250 mg load, then 100/50 mg q8h on hemodialysis days [5]. There is no increased clearance of C/T in patients with cystic fibrosis as seen with some drugs.

As with other β-lactam agents, C/T exhibit time-dependent killing of bacteria. The mean percentage of time needed above the MIC (%T > MIC) for 1-log kill for *P. aeruginosa* and wild-type *Enterobacteriaceae* was 31.5–31.6% and 34.8% for ESBL-producers [5].

### 3.2.3 Resistance to Ceftolozane/Tazobactam

*Enterobacteriaceae* can acquire resistance to C/T through acquisition of carbapenemases that are not inhibited by tazobactam (i.e., metallo-β-lactamases, KPC, GES-type enzymes). ESBL and AmpC producing gram-negative bacteria which are susceptible to C/T depend on the enzyme type and species [5]. Although the propensity for pseudomonas-acquired resistance to C/T appears lower than other agents (meropenem, ceftazidime, and ciprofloxacin) [6], acquired resistance has been reported from clinical isolates, with similar enzymes as found in resistant *Enterobacteriaceae*. However, *P. aeruginosa* resistance development requires multiple mutations with overexpression and modifications of AmpC [7]. In France, among 420 *P. aeruginosa* isolates nonsusceptible to ceftazidime and/or imipenem 42 (10%) were C/T resistant, and these resistant phenotypes were associated with extremely high cephalosporinase PDC [8]. In Portugal and Spain, C/T-resistant *P. aeruginosa* was significantly associated with GES-13 and VIM-type carbapenemase production [9].

### 3.2.4 Clinical Efficacy and Safety of Ceftolozane/Tazobactam

A phase 3 randomized controlled trial (RCT) has shown similar results with C/T plus metronidazole compared to meropenem for complicated intra-abdominal infections (cIAI) [10]. In a multicenter, phase 3 RCT of complicated urinary tract infections (cUTIs), including pyelonephritis, C/T was shown to be superior to levofloxacin in hospitalized patients for composite cure (microbiological and clinical) 5–9 days after treatment [11]. Both drugs were given for 7 days. This was largely due to levofloxacin-resistant uropathogens including *P. aeruginosa*, and, thus, levofloxacin may not have been an appropriate comparator, as quinolone resistant bacteria in complicated UTI are often present. Moreover, microbial cure for pyelonephritis is best assessed 1-month post-therapy.

Although C/T has high in vitro activity against drug-resistant *P. aeruginosa* and is suitable for treating these bacteria, these infections were underrepresented in the RCTs. Efficacy data has been limited to case series. In a recent retrospective, multicenter study of drug-resistant *P. aeruginosa* infections, the efficacy of C/T was compared to polymyxin or aminoglycoside in a cohort of 200 severely ill patients, 42% with severe sepsis and 52% with ventilator-associated pneumonia (VAP) [12]. Combination therapy was more commonly used with polymyxin/aminoglycoside (72% vs 15%). Although C/T therapy was independently associated with cure (adjusted odds ratio 2.63) and associated with significantly less acute kidney injury (AKI), there was no difference in hospital mortality. In a review of 128 cases of MDR-*P. aeruginosa* infections treated with C/T for non-approved indications, the overall clinical success rate was 76.2% [13]. Another retrospective review of 226 patients with MDR *P. aeruginosa* infections, 71.2% respiratory infections from 8 US medical centers, reported clinical failure rate of 37.6% and 30-day mortality of 17.3% [14]. New C/T MDR *P. aeruginosa* resistance was detected in 3 of 31 (9.7%) patients with follow-up cultures.

C/T was well tolerated in studies and post-marketing evaluations, similar to other cephalosporins, and higher doses up to 3000/1500 every 8 h were well tolerated. The most common side effects are nausea, vomiting, and diarrhea, and the incidence of *Clostridium difficile* infection is similar to other broad-spectrum cephalosporins [5]. One report of 48 evaluable patients treated with C/T found hypokalemia (4.2%) as the most common adverse event [13]. In the large retrospective cohort study of 226 patients, 9.7% experienced adverse effects including 9 acute kidney injury, 13 *C. difficile* infections, 1 hepatotoxicity, 2 encephalopathy, and 2 gastrointestinal intolerance [14].

### 3.2.5 *Summary and Place in Therapy of Ceftolozane/ Tazobactam*

Based on current data, C/T appears to have a special niche in the treatment of MDR-*P. aeruginosa* infections, but not MDR-*Enterobacteriaceae* especially ESBL producers. It has been posited that double the recommended dose may be appropriate for treatment of severe pseudomonas pneumonia (VAP) in order to achieve the probability of target attainment (PTA) >90% against *P. aeruginosa* with MIC up to 8 μg/ml [5].

## 3.3 Ceftazidime/Avibactam

The first-generation β-lactamase inhibitor (clavulanic acid, sulbactam, and tazobactam) efficacy has been eroded over the past two decades, but a new generation of non-β-lactam β-lactamase inhibitors has been created to counter this trend. Avibactam is a diazabicyclooctane β-lactamase inhibitor that is able to acylate the active site of serine β-lactamases reversibly, whereas the first-generation inhibitors lead to irreversible inhibition [15]. It is a potent inhibitor of class A β-lactamases, including the KPC enzymes, and class C enzymes. Ceftazidime/avibactam (Avycaz) combination was approved in the US in 2015 for the treatment of cUTIs and cIAIs (with metronidazole), whereas in the EU (marketed as Zavicefta) in addition it is approved for HAP (including VAP) and other infections with gram-negative bacteria with limited treatment options [16].

### 3.3.1 *In Vitro Activity of Ceftazidime/Avibactam*

Ceftazidime is an established anti-pseudomonas third-generation cephalosporin which binds to PBP to inhibit peptoglycan cross-linkage during cell wall synthesis to cause cell death, similar to other β-lactam agents. Avibactam has no significant intrinsic antimicrobial activity, but it protects ceftazidime from hydrolysis by a variety of serine enzymes. It inhibits a wide spectrum of class A β-lactamases including

TEM, SHV, CTX-M, PER, and KPC enzymes (including ESBL and some inhibitor-resistant enzymes), class C (i.e., AmpC), and certain class D β-lactamases (e.g., OXA-10, OXA-48). However, it does not inactivate class B enzymes or metallo-β-lactamases [16].

Ceftazidime/avibactam (CZA) has excellent in vitro activity against *Enterobacteriaceae*, 99.5% of 34,062 isolates susceptible with $MIC_{90}$ of 0.5 µg/ml from the INFORM global surveillance study in 2012–2014 [17]. These included

**Table 3.2** Comparative in vitro activity of ceftazidime/avibactam against *Enterobacteriaceae* and *P. aeruginosa*

| | | % Susceptible [$MIC_{90}$, µg/ml] | | | | |
|---|---|---|---|---|---|---|
| Organisms | No. of isolates | CZA | CZ | PTZ | MEM | AMK |
| *Enterobacteriaceae* | 34,062 | 99.5 [0.5] | 75.6 [64] | 84 [128] | 97.2 [0.12] | 96.3 [8] |
| ESBL/AmpC−+ | 5752 | 99.9 [0.5] | 16.1 [128] | 67.7 [>128] | 100 [0.06] | 93.3 [16] |
| KPC − + | 557 | 97.5 [4] | 3.9 [>128] | 0.9 [>128] | 3.1 [>8] | 48.3 [>32] |
| MEM-R | 961 | 83.5 [>128] | 5.3 [>128] | 3.2 [>128] | 0.0 [>8] | 58.1 [>32] |
| MBL − + | 145 | 3.4 [>128] | 2.1 [>128] | 6.9 [>128] | 0.0 [>8] | 59.3 [>32] |
| *E. coli* | 11,770 | 99.9 [0.25] | 78.9 [32] | 89.8 [32] | 99.6 [0.03] | 97.8 [8] |
| *K. pneumoniae* | 9098 | 99.0 [1] | 60.3 > 128] | 71.2 [>128] | 91.6 [0.25] | 93.0 [16] |
| *Enterobacter* spp. | 3931 | 98.8 [1] | 66.4 [128] | 74.9 [128] | 97.8 [0.12] | 97.8 [4] |
| *Citrobacter* spp. | 1889 | 99.3 [0.5] | 78.6 [128] | 84.5 [64] | 98.9 [0.06] | 98.9 [4] |
| *M. morganii* | 979 | 99.9 [0.12] | 88.0 [8] | 98.1 [2] | 100 [0.25] | 98.4 [8] |
| *P. vulgaris* | 995 | 99.9 [0.06] | 98.4 [0.12] | 99.7 [1] | 99.9 [0.12] | 99.2 [4] |
| *S. marcescens* | 784 | 99.2 [0.5] | 91.0 [4] | 93.4 [16] | 98.8 [0.12] | 94.8 [8] |
| *P. aeruginosa* | 7062 | 92.0 [8] | 77.0 [64] | 68.6 [>128] | 72.7 [>8] | 89.4 [32] |
| _KPC − + | 29 | 75.8 [32] | 0.0 [>128] | 0.0 [>128] | 0.0 [>8] | 75.9 [>32] |
| _CZ-R | 1627 | 65.4 [64] | 0.0 [>128] | 5.4 [>128] | 31.8 [>8] | 65.7 [>32] |
| _MEM-R | 1926 | 72.4 [64] | 42.4 [128] | 28.4 [>128] | 0.0 [>8] | 67.4 [>32] |

Data obtained from reference [13]

Abbreviations: *AMK* amikacin, *CZA* ceftazidime/avibactam, *CZ* ceftazidime, *MEM* meropenem, *PTZ* piperacillin/tazobactam, *CZ-R* ceftazidime resistant, *MEM-R* meropenem resistant

ESBL, AmpC-, and KPC-positive isolates, with 97.5–99.9% susceptible and 83.5% of carbapenem nonsusceptible isolates were susceptible to CZA [16]. Against *P. aeruginosa,* overall 92.0% of isolates from INFORM were susceptible with $MIC_{90}$ of 8 μg/ml, including 65% of ceftazidime-nonsusceptible and 72% of meropenem-nonsusceptible isolates. Table 3.2 summarizes the comparative in vitro activity of CZA.

Compared with C/T, CZA has similar susceptibility rates but lower MIC against ESBL-producing *Enterobacteriaceae* and higher MIC for *P. aeruginosa* [18, 19], but better in vitro activity against carbapenem-resistant *Enterobacteriaceae*, as tazobactam is inactive against AmpC β-lactamases, KPCs, and OXA-carbapenemases [16].

CZA has poor activity against most gram-positive bacteria and anaerobes, and *Acinetobacter* spp. and *S. maltophilia* are generally resistant. Acquired resistance appears to be low and 77.8% of nonsusceptible *Enterobacteriaceae* produced metallo-β-lactamases [16]. Acquired resistance is also due to production of β-lactamases not inactivated by avibactam, most commonly class B and D enzymes, but occasionally due to derepressed AmpC and KPC-3 mutations [20, 21]. Other resistant mechanisms besides β-lactamases include changes in drug target (mutant PBPs), decreased membrane permeability, active reflux, or combination of mechanisms.

### *3.3.2 Pharmacokinetic/Pharmacodynamics of Ceftazidime/Avibactam*

The pharmacokinetics of ceftazidime and avibactam are not affected when given together and both drugs are eliminated unchanged in the urine. The elimination half-lives are similar, 2.3 h for avibactam and 3.5 h for ceftazidime, with protein binding of about 8% and 10%, respectively [18]. A 2 h infusion of 2 g/0.5 g results in peak concentration of ceftazidime of 88.1 μg/ml and avibactam of 15.2 μg/ml; trough concentrations after multiple doses every 8 h after 10 days were 4.5 and 0.25 μg/ml [22]. The volumes of distribution are equivalent to the extracellular volume.

The recommended dose of CZA for CrCl ≥51 ml/min is 2.5 g (2/0.5 g) every 8 h infused over 2 h, adjusted according to degree of renal dysfunction: 25 g (1.0/0.25) every 8 h for CrCl 31–50 ml/min; 0.94 g (075/0.19 g) every 12 h for CrCl of 16–30; 0.94 g every 24 h for CrCl 6–15 ml/min; and for patients on hemodialysis, 0.94 g after dialysis on days of hemodialysis (pharmaceutical drug manual).

As with other β-lactam agents, the %T > MIC of CZA provide the best predictor for therapeutic response. A previous pharmacokinetic model found that for ceftazidime in patients with HAP, favorable outcome could be predicted with %T > MIC at >45% [23]. The pharmacokinetic/pharmacodynamic (PK/PD) index of avibactam appears best to correlate with the percentage of the dosing interval above a given threshold concentration (%T > $C_T$), 1 μg/ml in a ceftazidime-resistant neutropenic animal model [24]. Simulation models predicted that the current recommended dosage used in phase 3 trials would provide a joint PK/PD target of ceftazidime 50%T > MIC and avibactam 50%T > $C_{T\,to}$, which based on animal experiments should lead to >1–2 log bacterial (*P. aeruginosa*) killing [16].

### *3.3.3 Therapeutic Efficacy of Ceftazidime/Avibactam*

CZA was compared to doripenem for cUTI (including pyelonephritis) for 10–14 days, with option of oral therapy after ≥5 days intravenous therapy, in two double-blind, noninferiority RCTs (RECAPTURE-1 and RECAPTURE-2) [16]. ESBL-positive *Enterobacteriaceae* and ceftazidime-nonsusceptible pathogens were recovered in 19.1–19.6%. CZA was noninferior to doripenem and at late follow-up (45–52 days post-randomization), the microbiological cure rate was higher with CZA (68.2 vs 60.9%). In an open-label randomized trial (REPRISE), CZA was compared to best available therapy for patients with ceftazidime-nonsusceptible *Enterobacteriaceae* or *P. aeruginosa* (92% with cUTI and 8% with cIAI) in 333 patients [16]. Clinical cure rates were similar between the treatment groups at 21–25 days post-randomization.

In cIAI, three double-blind, phase 3 RCTs (RECLAIM-1, RECLAIM-2, and RECLAIM-3, the latter in Asia) were performed to compare CZA with metronidazole compared to meropenem for 5–14 days in 1507 patients [16]. CZA plus metronidazole was noninferior to meropenem (cure rates 82.5 vs 84.9%), and rates of cure for ceftazidime-susceptible compared to ceftazidime-nonsusceptible infections were similar.

The efficacy of CZA for HAP (including VAP) was assessed in an international, phase 3, double-blind, randomized REPROVE trial, compared with meropenem in 879 patients [16]. Treatment was for 7–14 days, despite the fact previous studies and guidelines recommend only 7-days therapy. CZA was noninferior to meropenem and cure rates and all-cause mortality at 28 days were similar (8.4% and 7.3%).

Analysis of pooled data from the 5 RCTs to assess the efficacy of CZA for MDR *Enterobacteriaceae* and *P. aeruginosa* ($n = 876$) found similar cure rates with comparator groups, 76.7% vs 69.0% for all *Enterobacteriaceae* and 71.0 vs 78.9% for *P. aeruginosa* [16].

Data on the effectiveness of CZA in critically ill patients with carbapenem-resistant bacteria are still limited even after its approval in 2015. In an observational cohort study of patients from two intensive care units (ICUs) on ventilators, CZA ($n = 41$) was compared with other available antibiotics ($n = 36$) for the treatment of CRE infections [25]. Clinical cure rate was found in 80.5% of patients treated with CZA versus 52.8% in patients treated with other antibiotics, $p = 0.01$; and the 28-day survival was greater with CZA (85.4% vs 61.1%, $p = 0.035$).

### 3.3.4 *Safety of Ceftazidime/Avibactam*

The safety and tolerability of CZA generally is similar to ceftazidime alone or other cephalosporins. The most common side effects from phase 11 and 111 trials were nausea, diarrhea, and positive Coombs test [16]. *C. difficile* colitis was reported in a few patients in these trials. Neurological complications (i.e., tremor, myoclonus, seizures, encephalopathy, and coma) were reported occasionally in patients with renal impairment with no dosage adjustment of CZA.

### 3.3.5 *Role of Ceftazidime/Avibactam in Gram-Negative Bacteria Infections*

CZA provides another choice for treatment of MDR gram-negative infections including cUTI, cIAI, HAP, and sepsis, but its niche in therapy compared to other new antibiotics is yet to be determined and will need further comparative trials between these newly approved agents. CZA appears to be a better choice than C/T for ESBL *Enterobacteriaceae*, but it is unknown whether its efficacy will be equal to or greater for MDR *P. aeruginosa*. It would not be suitable for metallo-β-lactamase producing carbapenem-resistant pathogens or *Acinetobacter* spp. infections.

## 3.4 Carbapenem/β-Lactamase Inhibitor Combinations

In the last two decades, the carbapenems have been the "reserved weapons" to treat MDR gram-negative bacterial infections, including ESBL organisms. However, with increasing use over the years, the emergence of carbapenem-resistant pathogens now poses a global threat. The carbapenem-resistant *Acinetobacter baumannii* (CRAB), carbapenem-resistant *Ps. aeruginosa* (CRPA), and carbapenem-resistant *Enterobacteriaceae* (CRE) are among the WHO list of antibiotic-resistant "priority pathogens" that poses a threat to the global health of the world's populations [26]. The new cephalosporin-β-lactamase inhibitors are unable to treat these pathogens reliably, which are often treated with polymyxins and tigecycline, but resistance to these antibiotics is increasing [27]. Thus, to meet this challenge existing carbapenems have been combined with novel β-lactamase inhibitors.

### 3.4.1 *Meropenem/Vaborbactam*

Meropenem/vaborbactam (M/V), marketed as Vabomere, was approved by the FDA in 2017 for treatment of cUTI including pyelonephritis and more recently in the EU for cUTI, cIAI, HAP, VAP, and infections with MDR gram-negative organisms with limited treatment options. Vaborbactam is a first-in-class boronic acid transitional state inhibitor (BATSI) that inhibits class A carbapenemases, especially *Klebsiella pneumoniae* carbapenemase (KPC) produced by *Enterobacteriaceae* [27], which is the most common carbapenemase found in the US. It is a broad-spectrum potent inhibitor of class A serine carbapenemases, including NMC-A, SME-2, CTX-M, SHV, and newly discovered BKC-1 and FRI-1, and class C β-lactamase (p99, MIR) [28]. Class B (e.g., NDM, VIM) and class D (e.g., OXA-48) carbapenemases are not inhibited by vaborbactam.

Vaborbactam has no antibacterial activity but combined with meropenem, it protects the agent from degradation by serine carbapenemase by a tight-binding reversible enzyme inhibition. It reduces the MIC of meropenem by ≥64-fold against bacterial strains producing class A serine carbapenemase. Vaborbactam and carbapenems crosses the bacterial outer membrane by using OmpK35 and OmpK36 porins, but unlike meropenem it is not a substrate for the MDR efflux pump AcrAB-To1C [28].

### 3.4.2 Antibacterial Activity of Meropenem/Vaborbactam

The FDA proposed susceptibility breakpoints for M/V against *Enterobacteriaceae* is ≤4/8 μg/ml by broth microdilution and > 99% of KPC-producing *Enterobacterales* were found susceptible in large surveillance studies [29]. M/V retains activity against KPC mutants resistant to CZA and is active against ESBL or AmpC-β-lactamase producing strains with permeability barrier. At least 90% or more *of K. oxytoca, E. aerogenes*, *Citrobacter freundii*, and *C. koseri* are susceptible to M/V. The activity of M/V is similar to meropenem for *P. aeruginosa* and *Acinetobacter* spp. and overall poor and similar for *S. maltophilia* and *Pandoraea* spp. [29]. Vaborbactam improves the activity over meropenem alone for MDR and extensively drug-resistant (EDR) *Enterobacteriaceae* and CRE [30]. M/V also has high activity against *Achromobacter* spp. and *Burkholderia* spp., which are found as respiratory pathogens mainly in patients with cystic fibrosis [28].

In a recent study, M/V and comparators in vitro activities were assessed against 152 CRE collected from US hospitals. M/V was active against 95.4% of the isolates (MIC ≤8 μ/ml) including all strains producing serine carbapenemases, but low activity against 7 isolates carrying metallo-β-lactamases and oxacillinases [31]. The most active comparator agents were tigecycline, colistin, and amikacin (63.2–96.7% susceptible).

### 3.4.3 Resistance to Meropenem/Vaborbactam

M/V resistance isolates are rarely found in surveillance studies. Resistant strains of clinical bacterial isolates may occur through coproduction of KPC and class B or D carbapenemases, or those with porin mutations and overexpression of efflux pumps

[32]. However, vaborbactam is less affected than avibactam by KPC-2 mutations that result in resistance to CZA [33]. The greatest reduction of M/V in vitro activity is noted in the presence of KPC-producing strains lacking OmpK35 and OmpK36 porins and overexpression of the multidrug efflux pump AcrAB [29]. Overexpression of KPC due to increased *bla*KPC gene copy number was also reported to lead to resistance selection [34]. Inactivation of both porins and OmpK36 has a greater effect in reducing the in vitro activity of M/V than inactivation of OmpK35 [29].

### 3.4.4 Pharmacologic/Pharmacodynamic Aspects of Meropenem/Vaborbactam

The recommended dosing of M/V is 4 g (2 g + 2 g) in 3 h infusion every 8 h, as clinical evidence indicate that meropenem (and other β-lactams) administered over prolonged infusion may result in improved clinical response and bacterial killing [35, 36]. The elimination half-life of meropenem (1–1.22 h) is similar to that of vaborbactam (1.68 h), and the peak plasma concentration and area under the plasma concentration curve show a dose-related linear increase [28, 31]. About 2% of meropenem and 33% of vaborbactam is plasma protein bound, and the volumes of distribution, 20.2 and 18.6 L, respectively, are related to the extracellular distribution [37]. Vaborbactam is not metabolized, and 75–90% of the drug is eliminated by the kidneys unchanged within 24–48 h, while meropenem undergoes 20–30% nonrenal elimination mainly due to metabolism of the parent compound by dipeptidases or by nonspecific degradation and 40–60% is recovered in the urine unchanged [28, 36]. Meropenem has good tissue penetration including the cerebrospinal fluid.

In patients with renal impairment, dose adjustment is needed for CrCL <50 ml/min as follows [32]:

- 2 g (1 g + 1 g) every 8 h in patients with CrCl 30–49 ml/min
- 2 g (1 g + 1 g) every 12 h in patients with CrCl 15–29 ml/min
- 1 g (0.5 g + 0.5 g) every 12 h in patients with CrCL <15 ml/min

The carbapenems, like all β-lactam agents, demonstrate time-dependent killing and the free drug concentration should be maintained above the MIC of the pathogen for prolonged periods at the site of infection—%T > MIC of >20% of the dosing time for bacteriostatic effect and > 40% for bactericidal effect [36]. In vitro and experimental animal studies suggest that a concentration of 8 mg/L of vaborbactam may be optimal for treatment of CRE [28]. PK-PD studies show that the current dosing of M/V produces sufficient concentration or exposure to produce bacterial killing and suppress the resistance of CRE [36].

### 3.4.5 Clinical Efficacy of Meropenem/Vaborbactam

The clinical efficacy and safety of M/V were evaluated in two phase 3 trials, Targeting Antibiotic Non-susceptible Gram-negative Organisms (TANGO) I and II. TANGO-I was a multicenter, noninferiority, RCT in adult patients ($n = 545$) with cUTI (including acute pyelonephritis) comparing M/V (2 g/2 g infused over 3 h every 8 h) to piperacillin/tazobactam (PTZ, 4 g/0.5 g infused over 30 min every 8 h) for 10 days [38]. Pyelonephritis was present in 59% and of the remaining patients 22% had removable source and 19% had non-removable source, and bacteremia occurred in 7%. After 15 doses of intravenous therapy, patients could be switched to oral levofloxacin (500 mg/day) to complete the 10 days. The overall success (composite of clinical cure and microbial eradication) was somewhat higher for M/V at test of cure (days 15–19), 74.5% vs 70.3%. The main limitations of this study were (i) 12% of the *Enterobacterales* were resistant to PTZ while almost all were meropenem susceptible and (ii) the study did not match for extended duration of infusion. M/V resistance was only found in one *Enterobacterales* and in 43% of *P. aeruginosa* [38].

TANGO-II was an open-label RCT comparing M/V to best available therapy (BAT) 2:1 for patients with different CRE infections (UTI, HAP/VAP, cIAI, and bacteremia) [39]. Seventy-seven patients with confirmed or suspected CRE infections were randomized, and 47 with confirmed CRE infections formed the primary analysis population. The dose and infusion time of M/V were the same as in TANGO-I for 7–14 days, and for BAT the regimens were mono/combination therapy with polymyxins, carbapenems, aminoglycosides, tigecycline, or CZA alone. Despite the small size, M/V was shown to be superior to BAT for clinical cure 7 days after treatment (59.4% vs 26.7%, $p = 0.02$); all-cause mortality at 28 days (15.6% vs 33.3%); and nephrotoxicity (4% vs 24%). The results of TANGO-II are supported by a recent prospective observational, single-site, study of 20 patients with CRE (70% in critical care) treated with M/V with clinical success and survival at 30 days of 65% and 90%, respectively [40]. Thirty-five percent had microbiological failure within 90 days and one patient had recurrent with M/V resistant, *omp*K36 mutant *K. pneumoniae*.

The safety and tolerability of M/V is similar to meropenem alone. In the TANGO-I trial, drug-related adverse events were 15.1% (2.6% severe events) for M/V and 12.8% (4.8% severe events) for PTZ. 2.6% of patients in the M/V group discontinued therapy because of adverse events compared to 5.1% with PTZ [36]. The most common adverse events for M/V compared to PTZ were headache (8.8% vs 4.4%), diarrhea (3.3% vs 4.4%), and infusion-related phlebitis (2.2% vs 0.7%); and severe adverse events included anemia and increased aspartate aminotransferase. In the TANGO-II trial, drug-related adverse events were lower for M/V compared to BAT (24.4% vs 44%) and event occurring >10% for M/V included diarrhea (1 associated with *C. difficile*), anemia, and hypokalemia [38]. Although meropenem has been associated with seizures, to date the addition of vaborbactam does not appear to increase the risk.

### 3.4.6 Imipenem-Cilastatin/Relebactam

Imipenem-cilastatin/relebactam (IMI/REL), named Recarbrio, was approved in the US in 2019 for treatment of cIAI and cUTI. Relebactam is a non-β-lactam, bicyclic diazabicyclooctane, β-lactamase inhibitor developed by building on the structure of avibactam by addition of a piperidine ring, which impedes the efflux from bacterial cells [40]. Similar to avibactam, it binds to the active site of serine β-lactamases of classes A and C, plus some class D (including carbapenemases); but unlike avibactam, relebactam is not degraded by desulfation [41].

Although carbapenem resistance is increasing globally, it is still low in North America and only 2.3% of *Enterobacteriaceae* infections in US hospitals are carbapenem resistant [42]. Carbapenem resistance is predominantly due to β-lactamases that hydrolyze the β-lactam ring such as serine carbapenemases (KPC, IMI), metallo-β-lactamases (VIM, NDM), and several of the OXA class, in addition mutations resulting in loss of porin channels and occasionally overexpression of efflux pumps [43]. Relebactam, similar to vaborbactam, does not inactivate metallo-β-lactamases or OXA-48-type carbapenemase [44]. Table 3.3 shows the activities of the β-lactamase inhibitors against different β-lactam enzymes [45].

**Table 3.3** Activities of the β-lactamase inhibitors in the new combinations on different β-lactamase enzymes

| Enzymes | Tazobactam | Avibactam | Vaborbactam | Relebactam |
|---|---|---|---|---|
| **Class A** | | | | |
| TEM | + | + | + | + |
| SHV | + | + | + | + |
| CTX-M | + | + | + | + |
| KPC | – | + | + | + |
| **Class B** | | | | |
| MBL | – | – | – | – |
| **Class C** | | | | |
| AmpC | – | + | + | + |
| **Class D** | | | | |
| OXA | – | ± | – | ± |

Data obtained from reference [39]

Note: TEM, SHV, and CTX-M are widely disseminated among the *Enterobacteriaceae* by plasmids to form the extended-spectrum phenotype (ESBL). KPC is the most common carbapenemase globally. AmpC organisms ["SPACE*]: Serratia, Pseudomonas or Proteus, Acinetobacter, Citrobacter,* and *Enterobacter.* OXA β-lactamases occur mainly in *Acinetobacte*r spp.

Abbreviations: *CTX-M* cefotaximase, *KPC Klebsiella pneumoniae* carbapenemase, *MBL* metallo-β-lactamase, *OXA* oxacillinase

### 3.4.7 In Vitro Activity of Imipenem/Relebactam

Relebactam greatly improves the activity of imipenem against most *Enterobacteriaceae* with lower MIC by 2- to 128-fold and against *P. aeruginosa* by eight-fold but does not improve the activity against *A. baumannii, S. maltophilia,* and most anaerobes [45]. This effect is similar to vaborbactam, which improves the activity of meropenem against most *Enterobacteriaceae* with 2-to >1000-fold MIC reduction, except it does not improve the activity against *P. aeruginosa,* as well as *A. baumannii* and *S. maltophilia.* Over 95% of *E. coli, K. pneumoniae, Citrobacter* spp., and *Enterobacter* spp. are susceptible to IMI/REL, but less with *Serratia marcescens* (87%), *Proteus mirabilis* (66%), and *Morganella morganii* (24%) [46]. For *Enterobacteriaceae*, the approved breakpoints for IMI/REL are susceptible ≤1/4 μg/ml, intermediate 2/4 μg/ml, and resistant ≥4/4 μg/ml and for *P. aeruginosa*, susceptible ≤2/4, intermediate 4/4 μg/ml, and resistant ≥8/4 μg/ml [47].

Among a selection of 106 CRE bloodstream isolates, only IMI/REL and CZA of 19 antimicrobial agents showed good in vitro activity against >90%, but against OXA-48-like CRE isolates only CZA and polymyxin/colistin were active against >90% [45]. In a large collection of *P. aeruginosa* clinical isolates ($n = 1445$), IMI/REL showed the highest susceptibility rate (97.3%), followed by colistin and C/T, both 94.6% [46]. IMI/REL remained active against extensive drug resistance (XDR), including strains that developed resistance to CZA and C/T. *P. aeruginosa* resistance to imipenem is by downregulation of porin protein synthesis in combination with AmpC overproduction and REL inhibit the latter to lower the MIC [48]. In contrast, meropenem-resistant *P. aeruginosa* is the result of impermeability and overexpression of efflux pumps rather than β-lactamase production.

There is no data so far on the development of resistance to IMI/REL after its use, but this predictably will occur with increased usage. This will likely occur from acquisition of β-lactamases not inhibited by relebactam (i.e., metallo-β-lactamases, or class D—OXA-48, OXA-51, OXA-23) and mutations affecting porins (*ompk* 35, *ompk* 36) and overexpression of efflux pumps in combination.

### 3.4.8 Pharmacokinetics/Pharmacodynamics of Imipenem/Relebactam

The pharmacokinetics of IMI and REL are similar and are unchanged with single dose or multiple doses. The protein binding of both are about 20%, the half-life of IMI is ≈1 h and REL 1.35 h, 60–70% of IMI is cleared by the kidneys versus 90% for REL, and the volume of distribution reflects the extracellular volume with similar values (16.1 L and 15.93 L); the peak concentration of IMI after 500 mg infusion is 35.6 μg/ml and that of REL is 16.8 μg/ml after 250 mg infusion in healthy volunteers [46, 49]. Co-administration of the two agents did not affect the pharmacokinetics of each other.

The PK/PD marker of efficacy of IMI/REL has been studied against *E. coli*, *K. pneumoniae*, *S. marcescens*, and *P. aeruginosa* in the hollow-fiber model. With the current dosing of 500 mg/250 mg (IMI/REL) every 6 h infused over 30 min, PK/PD simulations show that the drug exposures provide coverage of >90% of carbapenem-resistant strains [46]. This was based on PK/PD targets of 40% *f*T > MIC for IMI (corresponding to 2-$\log_{10}$ CFU reduction) and 7.5 *f*AUC/MIC for REL (corresponding to 2-$\log_{10}$ CFU reduction).

The current recommended dose of IMI/REL is 1.25 g (500 mg IMI, 500 mg cilastatin, 250 mg REL) every 6 h for normal renal function. For renal impairment dose adjustment is recommended: CrCl 60–89 ml/min, 1 g every 6 h (400 mg, 400 mg, 200 mg); CrCl of 30–59 ml/min, 0.75 g (300 mg, 300 mg, 150 mg) every 6 h; CrCl 15–29 ml/min, 0.5 g (200 mg, 200 mg, 100 mg) every 6 h; and end-stage renal failure on hemodialysis—0.5 g every 6 h (manufacture drug manual).

### 3.4.9 Clinical Efficacy and Side Effects of Imipenem/Relebactam

The efficacy and tolerability of IMI/REL were evaluated in two phase II, multicenter, double-blind RCTs, one for cUTI and the other for cIAI. IMI (500 mg)/REL [125 or 250 mg) versus IMI alone were used in both trials; as expected, they provided limited efficacy data with no difference in outcomes, but provided safety and PK data of the new agent REL, which did not increase the side effects combined with IMI [46].

Two phase III, multicenter, double-blind RCTs to evaluate the efficacy of IMI/REL in more resistant gram-negative infections were performed [46]. RESTOR-IMI 1 only included 47 patients with IMI-resistant gram-negative provided descriptive comparison of efficacy versus colistin + IMI (500 mg q6h) for cUTI, cIAI, and HAP/VAP. *P. aeruginosa* was the most common infective organism in 77% of the cases. The clinical efficacy were similar between the two groups (71% vs 70%), but favorable response was higher with IMI/REL compared to IMI + colistin (81% vs 63%) for patients infected with *P. aeruginosa* [44]. RESTORE-IMI 2 compared IMI/REL to piperacillin-tazobactam (PTZ) in 537 patients with HAP/VAP for 7–14 days [50]. The most common pathogens were *K. pneumoniae* (25.6%) and *P. aeruginosa* (18.9%). IMI/REL was noninferior to PTZ for day 28 all-cause mortality (15.9% for IMI/REL vs 21.3% for PTZ) and favorable clinical response—61.0% vs 55.8%, respectively.

IMI/REL adverse effects appear similar to IMI alone. In phase II trials, the common drug-related adverse events were diarrhea, nausea, headache, and increased transaminases (1.4%). In the RESTORE IMI 2 trial drug-related adverse events occurred in 11.7% of the IMI/REL arm and 9.7% of the PTZ arm, with adverse events leading to discontinuation in 5.6% and 8.2%, respectively [50].

### 3.4.10 Place in Therapy of Meropenem/Vaborbactam and Imipenem/Relebactam

These two new carbapenem/β-lactamase inhibitors (M/V and IMI-REL) provide additional stockpile of our armaments to combat MDR gram-negative bacteria in addition to the new cephalosporin/β-lactamase inhibitors and others. These agents should be reserved for proven or suspected MDR gram-negative bacteria (i.e., CRE) such as in cUTI, cIAI, HAP/VAP, and hospitalized bacteremia. Although it may be difficult, comparative studies should be done with these new combinations for infection with likely or proven MDR gram-negative bacteria. Based on limited data, M/V may be better for CRE infections as it appears superior to the best available therapy, whereas IMI-REL results were similar to the comparators. However, IMI-REL may be better for some MDR *P. aeruginosa* as vaborbactam did not improve the activity of meropenem against resistant strains in vitro. Recent guidelines list IMI-REL but not M/V as one of the options for treatment drug-treatment resistant *P. aeruginosa* infections of complicated UTI and infection outside the urinary tract, but both agents are listed for CRE infections [51].

## References

1. Drawz SM, Bonomo RA. Three decades of β-lactamase inhibitors. Clin Microbiol Rev. 2010;23:160–201.
2. Zhanel GG, Chung P, Adam H, et al. Ceftolozane/tazobactam: a novel cephalosporin/β--lactamase inhibitor combination with activity against multidrug-resistant gram-negative bacilli. Drugs. 2014;74:31–51.
3. Moya B, Zamorano L, Juan C, et al. Affinity of the new cephalosporin CXA-101 to penicillin-binding proteins of *Pseudomonas aeruginosa*. Antimicrob Agents Chemother. 2010;54:3933–7.
4. Sebaaly J, Woods JA, Wargo KA. A review of ceftolozane/tazobactam for the treatment of infections caused by multidrug-resistant pathogens. Infect Dis Clin Pract. 2018;26:198–203.
5. Giacobbe DR, Bassetti M, De Rosa FG, et al. Ceftolozane/tazobactam: place in therapy. Expert Rev Anti-Infect Ther. 2018;16:307–20.
6. Livermore DM, Mushtaq S, Meunier D, et al. Activity of ceftolozane/tazobactam against surveillance and 'problem' Enterobacteriaceae, *Pseudomonas aeruginosa* and non-fermenters from the British Isles. J Antimicrob Chemother. 2017;72:2278–9.
7. Cabot G, Bruchmann S, Mulet X, et al. *Pseudomonas aeruginosa* ceftolozane-tazobactam resistance development requires multiple mutations leading to overexpression and structural modifications of AmpC. Antimicrob Agents Chemothetr. 2014;58:3091–9.
8. Fournier D, Carrier R, Bour M, et al. Mechanisms of resistance to ceftolozane/tazobactam in *Pseudomonas aeruginosa:* results of the GERPA multicenter study. Antimicrob Agents Chemother. 2021;65(2):e01117–20. https://doi.org/10.1128/AAC.01117-20.
9. Hernandez-Garcia M, Garcio-Castillo M, Garcia-Fernandez S, et al. Distinct epidemiology and resistant mechanisms affecting ceftolozane/tazobactam in *Pseudomonas aeruginosa* isolates recovered from ICU patients in Spain and Portugal depicts WGS. J Antimicrob Chemother. 2021;76:370–9.

10. Solomkin J, Hershberger E, Miller B, et al. Ceftolozane/tazobactam plus metronidazole for complicated intra-abdominal infections in an era of multidrug resistance: results from a randomized, double-blind, phase 3 trial [ASPECT-cIAI]. Clin Infect Dis. 2015;60:1462–71.
11. Wagenlehner FM, Umeh O, Steenbergen J, et al. Ceftolozane/tazobactam compared to levofloxacin in the treatment of complicated urinary tract infections, including pyelonephritis: a randomized, double-blind, phase 3 trial [ASPECT-cUTI]. Lancet. 2015;385:1949–56.
12. Pogue JM, Kaye KS, Veve MP, et al. Ceftolozane/tazobactam vs polymyxin or aminoglycoside-based regimens for the treatment of drug-resistant *Pseudomonas aeruginosa*. Clin Infect Dis. 2020;71:304–10.
13. Maraolo AE, Mazzitelli M, Trecarichi EM, Buonomo AR, Torti C, Gentile I. Ceftolozane/tazobactam for difficult-to-treat *Pseudomonas aeruginosa* infections: a systematic review of its efficacy and safety for off-label indications. Int J Antimicrob Agents. 2020;55:105891. https://doi.org/10.1016/j.ijantimicag.2020.105891.
14. Jorgensen SC, Trinh TD, Zasowski EJ, et al. Real-world experience with ceftolozane-tazobactam for multidrug-resistant gram negative bacterial infections. Antimicrob Agents Chemother. 2020;64:e02291–19. https://doi.org/10.1128/AAC.02291-19.
15. Tooke CL, Hinchcliffe P, Braggington EC, Colenso CK, Hirvonen VHA, Takebayashi Y, Spencer J. β-lactamases and β-lactamase inhibitors in the 21st century. J Mol Biol. 2019;431:3472–500.
16. Shirley m. Ceftazidime-avibactam: a review in the treatment of serious gram-negative bacterial infections. Drugs. 2018;78:675–92.
17. Karlowsky JA, Buiedenbach DJ, Karmierczak KM, et al. Activity of ceftazidime-avibactam against extended spectrum- and AmpC β-lactamase-producing Enterobacteriaceae collected in the INFORM global surveillance study from 2012–2014. Antimicrob Agents Chemother. 2016;60:2849–57.
18. Buehrle DJ, Shields RK, Chen L, et al. Evaluation of the in vitro activity of ceftazidime-avibactam and ceftolozane-tazobactam against meropenem-resistant *Pseudomonas aeruginosa* isolates. Antimicrob Agents Chemother. 2016;60:3227–31.
19. Alatoom A, Esayed H, Lawlor K, et al. Comparison of antimicrobial activity between ceftolozane-tazobactam and ceftazidime-avibactam against multidrug-resistant isolates of *Escherichia coli, Klebsiella pneumonia, and Pseudomonas aeruginosa*. Int J Infect Dis. 2017;62:39–43.
20. Lahiri S, Walkup GK, Whittaker JD, et al. Selection and molecular characterization of ceftazidime/avibactam-resistant mutations in *Pseudomonas aeruginosa* strains containing derepressed AmpC. J Antimicrob Chemother. 2015;70:1650–8.
21. Haider G, Clancy CJ, Shields RK, et al. Mutations in $bla_{KPC-3}$ that confer ceftazidime-avibactam resistance encode novel KPC-3 variants that function as extended-spectrum β-lactamases. Antimicrob Agents Chemother. 2017;61:e02534–16.
22. Merdjian H, Rangaraju M, Tyarmal A. Safety and pharmacokinetics of single and multiple ascending doses of avibactam alone and in combination with ceftazidime in healthy male volunteers: results of two randomized, placebo-controlled studies. Clin Drug Investig. 2015;35:307–17.
23. Mulleer AE, Punt N, Mouton JW. Optimal exposures of ceftazidime predict the probability of microbiological and clinical outcome in the treatment of nosocomial pneumonia. J Antimcrob Chemother. 2013;68:900–6.
24. Berkhout J, Melchers MJ, van Mil AC, val. Pharmacokinetics of ceftazidime and avibactam in neutropenic mice with thigh or lung infection. Antimicrob Agents Chemother. 2016;60:368–75.
25. Tsolaki V, Mantzarlis K, Mpakalis A, et al. Ceftazidime-avibactam to treat life-threatening infections by carbapenem-resistant pathogens in critically ill mechanically ventilated patients. Antimicrob Agents Chemother. 2020;64:e02320–19. https://doi.org/10.1128/AAC.02320-19.
26. Willyard C. The drug-resistant bacteria that pose the greatest health threats. Nature. 2017;553:15.

27. Rodriguez-Bano J, Gutierrez-Gutierrez B, Machuca I, Pascual A. Treatment of infections caused by extended-spectrum-beta-lactamase-, AmpC-, and carbapenase-producing *Enterobacteriaceae*. Clin Microbiol Rev. 2018;31:e00079–17.
28. Lomovskaya O, Sun D, Rubino-Aparicio D, et al. Vaborbactam: spectrum of beta-lactamase inhibition and impact of resistance mechanisms on activity in Enterobacteriaceae. Antimicrob Agents Chemother. 2017;61:e01443–17.
29. Novelli A, Giacomo PD, Rossolini GM, Tumbarello M. Meropenem/vaborbactam: a next generation β-lactam β-lactamase inhibitor combination. Expert Rev Anti-Infect Ther. 2020;18:643–55.
30. Castanheira M, Huband MD, Mendes RE, Flamm RK. Meropenem-vaborbactam tested against contemporary gram-negative isolates collected worldwide during 2014, including carbapenem-resistant, KPC-producing, multi-drug resistant, and extensively drug-resistant Enterobacteriaceae. Antimicrob Agents Chemother. 2017;61:e00567–17.
31. Castanheira M, Doyle TB, Kantro V, Mendes RE, Shortridge D. Meropenem-vaborbactam activity against carbapenem-resistant *Enterobacterales* isolates collected in U.S. hospitals during 2016 to 2018. Antimicrob Agents Chemother. 2020;64:e01951–19. https://doi.org/10.1128/AAC.01951-19.
32. Dhillon S. Meropenem/vaborbactam: a review in complicated urinary tract infections. Drugs. 2018;78:1259–70.
33. Tsikovski R, Lomovskaya O. Potency of vaborbactam is less affected than that of avibactam in strains producing KPC-2 mutations that confer resistance to ceftazidime-avibactam. Antimicrob Agents Chemother. 2020;64:e01936–19. https://doi.org/10.1128/AAC.01936-19.
34. Sun D, Rubio-Aparicio D, Neson K, et al. Meropenem-vaborbactam resistance selection, resistance prevention, and molecular mechanisms in mutants of KPC-producing *Klebsiella pneumoniae*. Antimicrob Agents Chemother. 2017;61:e01694–17.
35. Grupper M, Kutij L, Nicolau DP. Continuous and prolonged intravenous β-lactam dosing: implications for the clinical laboratory. Clin Microbiol Rev. 2016;29:759–72.
36. Adembri C, Noevelli A. Pharmacokinetic and pharmacodynamic parameters of antimicrobials: potentials for providing dosing regimens that are less vulnerable to resistance. Clin Pharmacokinet. 2009;48:517–28.
37. Griffith DC, Sabet M, Tarazi Z, Lomovskaya O, Dudley MN. Pharmacokinetics/pharmacodynamics of vaborbactam, a novel beta-lactamase inhibitor, in combination with meropenem. Antimicrob Agents Chemother. 2019;63:e01659–18.
38. Kaye KS, Bhowmick T, Metallidis S, et al. Effect of meropenem-vaborbactam vs piperacillin-tazobactam on clinical cure or improvement and microbial eradication in complicated urinary tract infection: the TANGO I randomized clinical trial. JAMA. 2018;319:788–99.
39. Wunderlink RG, Giamarellos-Bourboulis EJ, Rahav G, et al. Effect and safety of meropenem-vaborbactam versus best-available therapy in patients with carbapenem-resistant *Enterobacteriaceae* infections: the TANGO II randomized clinical trial. Infect Dis Ther. 2018;7:439–55.
40. Sheilds RK, McCreary EK, Marini RV, et al. Early experience with meropenem-vaborbactam for treatment of carbapenem-resistant Enterobacteriaceae infections. Clin Infect Dis. 2020;71:667–71.
41. Blizzard TA, Chen H, Kim S, et al. Discovery of MK-7655. A beta-lactamase inhibitor for combination with Primaxin [R]. Bioorg Med Chem Lett. 2014;3:439–55.
42. Lodise T, Ye MJ, Zhan Q. Prevalence of invasive infections due to carbapenem-resistant Enterobacteriaceae among adult patients in U.S. hospitals. Antimicrob Agents Chemother. 2017;61:e00228–17.
43. Meletis G. Carbapenem resistance: overview of the problem and future perspectives. Ther Adv Infect Dis. 2016;1:15–21.
44. Canver MC, Satlin MJ, Westblade LF, et al. Activity of imipenem-relebactam and comparator agents against genetically characterized isolates of carbapenem-resistant Enterobacteriaceae. Antimicrob Agents Chemother. 2019;63:e00672–19.

45. Zhanel GG, Lawrence CK, Adam H, et al. Imipenem-relebactam and meropenem-vaborbactam: two novel carbapenem-β-lactamase inhibitor combinations. Drugs. 2018;78:65–98.
46. Smith JR, Rybak J, Claceys KC. Imipenem-cilastatin-relebactam: a novel β-lactamase inhibitor combination for the treatment of multidrug-resistant gram-negative infections. Pharmacotherapy. 2020;40:343–56.
47. Fraile-Ribot PA, Zamorano L, Orellana R, et al. Activity of imipenem-relebactam against a large collection of *Pseudomonas aeruginosa* clinical isolates and isogenic β-lactam-resistant mutants. Antimicrob Agents Chemother. 2020;64:e02165–19.
48. Livermore DM, Warner M, Mushtaq S. Activity of MK-7655 combined with imipenem against *Enterobacteriaceae* and *Pseudomonas aeruginosa*. J Antimicrob Chemother. 2013;10:2286–90.
49. Lucasti C, Vasile L, Sandesc D, et al. Phase 2, dose-ranging study of relebactam with imipenem-cilastatin in subjects with complicated intra-abdominal infection. Animicrob Agents Chemother. 2016;60:6234–43.
50. Titov I, Wunderink RG, Roquilly A, et al. A randomized, double-blind, multicenter trial comparing efficacy and safety of imipenem/cilastatin/relebactam versus piperacillin/tazobactam in adults with hospital acquired or ventilator-associated bacterial pneumonia [RESTORE-IMI 2 study]. Clin Infect Dis. 2020;73:ciaa803. https://doi.org/10.1093/cid/ciaa803.
51. Tammma PD, Aitken SL, Bonomo RA, Mathers AJ, van Duin D, Clancy C. Infectious Diseases Society of America 2022 guidance on the treatment of extended-spectrum β-lactamase producing Enterobacterales [ESBL-E], carbapenem-resistant Enterobacterales [CRE], and *Pseudomonas aeruginosa* with difficult-to-treat resistance [DTR-*P. aeruginosa*]. Clin Infect Dis. 2022;75:187–212.

# Chapter 4
# New Glycopeptides: Telavancin, Dalbavancin, and Oritavancin

## 4.1 Introduction

Glycopeptide antibiotics are glycosylated tricyclic or tetracyclic heptapeptides derived from soil actinomycetes used to treat serious gram-positive bacterial pathogens resistant to β-lactam agents. Vancomycin was the first of its class discovered in 1950 and introduced for clinical use in 1958 but was rarely used until the 1980s when methicillin-resistant *Staphylococcus aureus* (MRSA) started to proliferate in hospitals [1]. Vancomycin still remains the first-line therapy for serious MRSA infection, which is the most common antibiotic-resistant bacteria found in hospitals and is now widespread in the community. The sparse use of vancomycin for the first 30 years after marketing most likely contributed to low level of vancomycin-resistant bacteria present to date. High level resistance to vancomycin was first reported in enterococci (VRE) in 1988 and spreading since then [1]. In 2017 the Centers for Disease Control and Prevention (CDC) in the United States (US) estimated there were 54,500 hospitalized patients infected with VRE associated with 5400 estimated deaths (CDC: Antibiotic Resistance Threats in the United States, 2019). MRSA with intermediate susceptibility to vancomycin (VISA, MIC 3–8 μg/ml) was reported in 1997 and high level resistance (VRSA, MIC≥16/μg/ml) was noted in the US in 2002 [1]. The CDC estimated that there were 323,700 hospitalized patients infected with MRSA with 10,600 associated deaths in 2017 in the US.

The other first-generation naturally occurring glycopeptide produced by actinomycetes is teicoplanin, reported in 1978 and marketed for clinical use in Europe in 1988 and in Japan in 1998, but was never approved in the US [1, 2]. Vancomycin and teicoplanin share the core heptapeptide scaffold; while vancomycin has five aromatic and two aliphatic residues, teicoplanin has seven aromatic residues [2]. Teicoplanin has longer half-life and better safety profile than vancomycin with rare instances of red man syndrome (does not cause histamine release), ototoxicity. and nephrotoxicity [2].



I. W. Fong, *New Antimicrobials: For the Present and the Future*, Emerging Infectious Diseases of the 21st Century,
https://doi.org/10.1007/978-3-031-26078-0_4

The second-generation glycopeptides (telavancin, dalbavancin, and oritavancin) are semisynthetic derivatives of the natural compound with greater potency and better pharmacokinetic properties than vancomycin [3].

## 4.2 Mechanism of Action and Resistance Mechanism

The glycopeptide antibiotics inhibit the bacterial wall synthesis by binding to the membrane-bound lipid II precursor of peptidoglycan (the D-Ala-D-Ala dipeptide terminus), destabilizing the integrity of the cell wall with cell death [1, 2]. Gram-negative bacteria are protected by the outer lipopolysaccharide membrane impermeable to large molecules. Glycopeptide-resistant genes (called *van*) exist before the discovery of antibiotics and were found in ancient DNA from 30,000-year-old permafrost [1]. Glycopeptide resistance mutations occur, mainly in enterococci, through remodeling of the lipid II precursor to the D-Ala-D-Ala terminal peptide to D-Ala-D-Lac (*vanA, vanB, vanD*) or less commonly D-Ala-D-Ser (*vanC, vanE, vanG*) phenotype [1]. *VanA*-type is associated with high level inducible resistance to both vancomycin and teicoplanin, mediated by transposable genes (i.e., *Tn*1546) [2]. VISA resistance is due to excess accumulation of peptidoglycan to produce thickened cell wall, and the VRSA resistance is similar to VRE *vanA* modification of lipid II [1].

## 4.3 Second-Generation Glycopeptides

The second-generation glycopeptides are semisynthetic derivatives of the natural products and include telavancin, dalbavancin, and oritavancin and are considered as lipoglycopeptides. These new agents differ from vancomycin by the presence of a lipophilic side chain.

### *4.3.1 Telavancin*

Telavancin (Vibativ) was introduced in 2009 and approved in the US initially for complicated skin and skin structure infections (cSSSIs) due to gram-positive bacteria and then later in the US and Europe for hospital-acquired and ventilator-associated pneumonia (HAP/VAP) caused by *S. aureus* in 2013. Telavancin is derived from the modification of vancomycin structure by the addition of hydrophobic and hydrophilic side chains [1]. The lipophilic component increases the membrane interaction to produce greater potency against gram-positive bacteria, while the hydrophilic group improves the pharmacologic properties (promote tissue distribution and clearance) and reduce the nephrotoxic effects [4]. Besides inhibiting cell wall synthesis, it disrupts bacterial cell membrane and alters the cell permeability, resulting in its rapid bactericidal activity [5].

### 4.3.2 Pharmacology/Pharmacokinetics of Telavancin

Telavancin has a half-life of about 8 h and can be given once daily, and it is mainly excreted by the kidneys with 60–70% excreted unchanged in the urine [6]. However, it has higher protein binding than vancomycin (93% versus 50%), but the presence of serum or albumin only increases its MIC two-fold for *Staphylococcus* spp. [6], thus indicating that it is weakly protein bound. Due to its high protein binding, telavancin has low volume of distribution of 11 L (115 ml/kg) but penetrates adequately into blister fluid, lung epithelium, and within macrophages [7]. Despite its low penetration of the inflamed meninges in rabbits (2%), it sterilized the cerebrospinal fluid (CSF) of penicillin-resistant *Streptococcus pneumoniae* in 6 of 10 animals [7].

Pharmacodynamic studies showed that its bactericidal activity is concentration dependent [8], with a post-antibiotic effect (PAE) of 4–6 h against MRSA [6]. However, vancomycin is slowly bactericidal with 3 $\log_{10}$ decrease of colony-forming units (cfu) in 24 h and the PAE is only 1 h [6], as the bactericidal activity is not concentration dependent. Animal model and other studies indicate that the parameter that best predicts telavancin activity is the AUC/MIC (area under the concentration curve/minimum inhibitory concentration) ratio; higher ratio resulted in greater killing and longer delay in regrowth of the bacteria. The recommended dose of 10 mg/kg once daily results in a ratio of 50 and minimum plasma concentration of 5 mg/L, the lowest concentration that prevented bacterial growth at 24 h [9]; and maximal bactericidal activity was found at ratio of AUC/MIC of 404 [6]. Vancomycin efficacy is also best predicted by AUC/MIC of 400–600 mg.hr./L [10].

The recommended dose in patients with creatinine clearance (CrCl) of >50 ml/min is 10 mg/kg infused over 1 h daily, for CrCl 30–50 ml/min, 7.5 mg/kg every 24 h, and for 10- > 30 ml/min 10 mg/kg every 48 h (Global $R_X$Ph). The same dose can be given for patients with complete renal failure on hemodialysis without supplemental dose after dialysis.

### 4.3.3 In Vitro Activity of Telavancin

Telavancin is active against gram-positive aerobic and anaerobic bacteria with a similar spectrum as vancomycin. However, it is rapidly bactericidal and more potent than vancomycin with MICs two to eight times lower. The microbial activity is similar for telavancin, oritavancin, and dalbavancin against *Staphylococcus* spp., *Enterococcus* spp., *Streptococcus* spp., *Clostridium* spp., *Corynebacterium* spp., *Actinomyces* spp., *Lactobacillus* spp., and *Propionibacterium* spp. (see Table 4.1). For *S. aureus* including MRSA, the telavancin $MIC_{90}$ is 0.5 μg/ml or less and for

**Table 4.1** Comparative susceptibility of new glycopeptides and older agents ($MIC_{90}$) (μg/ml)

| Organisms | Vancomycin | Daptomycin | Telavancin | Dalbavancin | Oritavancin |
|---|---|---|---|---|---|
| MSSA | 1–2 | 0.5 | 0.25–0.5 | 0.06 | 0.12 |
| MRSA | 1–2 | 0.5–1 | 0.5 | 0.06 | 0.25 |
| CoNS | 2–4 | 0.5–1 | 0.5–1 | 0.06 | 0.5 |
| *S. pyogenes* | 0.5–1 | 0.25–0.6 | 0.03–0.06 | 0.06 | 0.25 |
| S. *agalactiae* | 0.5 | 0.25–1 | 0.06–0.125 | 0.12 | 0.12. |
| *S. pneumoniae* | 0.5 | 0.12 | 0.03 | 0.03 | 0.004 |
| *E. faecalis* (VS) | 2 | 1 | 0.5 | 0.06 | 0.06 |
| *E. faecalis* (VR) | 512 | 1–2 | 8 | 32 | 1 |
| *E. faecium* (VS) | 0.5–1 | 4 | 0.25–0.5 | 0.12 | 0.015 |
| *E. faecium* (VanA) | 512 | 4 | 8 | 32 | 0.25 |
| *E. faecium* (VanB) | 64 | 4 | 2 | 0.12 | 0.03 |
| *Clostridium* spp. | 1 | NA | 0.25 | 2 | 1 |

Data obtained from Refs. [6, 9]
Abbreviations: *CoNS* coagulase negative *Staphylococcus*, *MSSA* methicillin-sensitive *Staphylococcus aureus*, *MRSA* methicillin-resistant *Staphylococcus aureus*, *VS* vancomycin-susceptible, *VR* vancomycin-resistant

*S. epidermidis* 0.25–0.5 μg/ml for vancomycin-susceptible strains and 0.5–1.0 μg. ml for resistant strains [6]. The $MIC_{90}$ for *S. pneumoniae*, even for penicillin-resistant strains, is 0.03 μg/ml, and for β-hemolytic streptococci and *Streptococcus viridans*, the $MIC_{90}$ ranges from 0.06 to 0.125 μg/ml [6]. Against vancomycin-susceptible *E. faecalis* and *E. faecium,* the MICs range from ≤0.0015–0.5 and 0.06–2 μg/ml, respectively; but for VRE (*E. faecalis*) the MIC is 0.02–16 μg/ml and *E. faecium* is 0.015–16 μg/ml; the $MIC_{90}$ is 64 times lower than the $MIC_{90}$ of vancomycin [11].

Resistance to telavancin is similar to teicoplanin; it retains activity against strains that express *vanB,* but not *vanA* which is induced by the drug [4]. It has a high threshold for resistance development among staphylococci and enterococci and high level resistance in MRSA or VRE was not observed during in vitro resistance selection studies [1]. Selection of resistance in patients is rarely reported, but a three-fold increase in telavancin MIC was reported during the treatment of a patient with persistent bacteremia and mediastinitis with MRSA that evolved to VISA during treatment with vancomycin and daptomycin [12].

### *4.3.4 Clinical Efficacy and Side Effects of Telavancin*

The clinical efficacy and adverse effects of telavancin were assessed in patients with complicated skin and skin structure infections (cSSSI) due to gram-positive bacteria and in hospital-acquired pneumonia (HAP).

Two phase II randomized controlled, double-blind, multicenter trials (RCT) evaluated telavancin for cSSTI in the US and South Africa. The FAST I trial randomized 167 patients in a 1:1 ratio to receive either intravenous (IV) telavancin 7.5 mg/kg or standard therapy of either (i) vancomycin 1 g every 12 h or (ii) nafcillin or oxacillin 2 g every 6 h, (iii) or cloxacillin 0.5–1 g every 6 h for 4–14 days [13]. The cure rates between telavancin and standard therapy were similar 79% versus 80%, respectively. When patients with MRSA infections were analyzed, telavancin produced higher cure rate (82%, 18/22) than standard therapy (69%, 18/26), but not statistically significant. Serious adverse events were less in the telavancin group (4 vs 7), but the rate of discontinuation due to side effects was the same, 5%.

FAST II was also of similar design with the same subgroups but the dose of telavancin was 10 mg/kg once daily in a total of 195 patients for both groups [14]. The cure rates were similar for the two groups, but microbiological eradication for MRSA was greater with telavancin, 92% vs 68%, $p = 0.04$. Rates of serious reactions and discontinuation of medications due to adverse reactions were also similar (6% for telavancin and 3% for standard therapy).

Two phase III noninferiority RCTs were also completed in patients with cSSTIs (ATLAS-1 and ATLAS-2) for gram-positive pathogens. The combined trials recruited 1867 patients and 80% were evaluable for treatment with IV telavancin 10 mg/kg/day or IV vancomycin 1 g every 12 h for 7–14 days [15]. The clinical cure rate and microbiological eradication rates were similar between the groups, 88.6% (telavancin) and 86.2% (vancomycin). The results were similar for patients with MRSA infections. Serious adverse events occurred in 7% on telavancin and 4% on vancomycin, and mild adverse effect, such as nausea and vomiting, were common in both groups, 79% for telavancin and 72% for vancomycin [15].

Patients with HAP ($n = 1503$), including ventilator-associated pneumonia (VAP), were assessed in two phase III RCTs (ATTAIN trials) to compare telavancin 10 mg/kg/day to vancomycin 1 g every 12 h for 7–21 days [16]. Patients with MRSA HAP had a higher cure rate with telavancin (82%) than vancomycin (74%), but not statistically significant. This trend was evident in severely ill patients with APACHE II scores of ≥20 or patients over the age of 65 years. The safety profile of telavancin was similar to vancomycin; except in patients with moderate to severe renal impairment, there was increased mortality. Increased creatinine levels were more frequent with telavancin than vancomycin (16% vs 10%) [16].

### *4.3.5 Dalbavancin*

Dalbavancin (Dalvance) is a semisynthetic drug derived from a teicoplanin-like natural glycopeptide produced by an actinomycete species and was approved for clinical use in 2014 for bacterial skin and softy tissue infections in the US. Amidation of the C-terminal carboxyl group with a dimethylaminopropylamine group increases its activity against staphylococci, and the long lipophilic side chain results in extended half-life, which allows once weekly dosing [9]. Like all glycopeptides, it inhibits cell wall synthesis by binding to the terminal D-Ala-D-Ala in the peptidoglycan chains hindering polymerization and cross-linking resulting in destabilization of the cell wall and bacterial cell death. The lipophilic side chain enhances its activity by allowing greater binding to its target site and membrane anchoring [9].

### 4.3.6 Pharmacokinetics and Pharmacodynamics of Dalbavancin

Dalbavancin has linear pharmacokinetic properties and is highly protein bound, 93–98%, with a prolonged terminal half-life of 147–258 h [9]. It is eliminated by both renal and non-renal mechanisms with only 42% of the dose excreted unchanged in the urine and 12% removed as hydroxyl-dalbavancin. Although in animal studies about 50% of the unchanged drug is excreted in bile, in humans only 20% was found in the feces [17]. Despite the high protein binding, the drug is widely distributed in the body and the concentration in blister fluid is higher than telavancin and oritavancin (60% vs 40% and 19%, respectively) [9]. In healthy adults after 1 g dalbavancin IV, the peak serum concentration was 278.3–301 µg/ml and the mean AUC was 23,843 µg.h/ml (see Table 4.2 for the pharmacokinetic parameters for the glycopeptides [9]).

The standard dose of dalbavancin is 1000 mg infused over 30 min. on day 1 and 500 mg 8 days later if needed. Recently a single dose of 1500 mg has also been approved by the FDA for the same indication [18]. No dose adjustment is needed for mild to moderate renal impairment, but for severe renal impairment, CrCl <30 ml/min, the two-dose regimen is 750 mg on day 1 and 375 mg on day 8, or the single dose of 1125 mg [17].

Dalbavancin, similar to telavancin and oritavancin, exhibits concentration-dependent bactericidal activity with increased activity with increasing concentration, while vancomycin demonstrated time-dependent killing [9]. In animal models

**Table 4.2** Pharmacokinetics of glycopeptides in human volunteers at recommended doses

| Parameter | Vancomycin(15 mg/kg bid) | Telavancin (10 mg/kg od) | Dalbavancin (1 g day 1, 500 mg day 8) | Oritavancin (1200 mg od) |
|---|---|---|---|---|
| $C_{max}$ (µ/ml) | 20–50 | 88 | 312 | 138 |
| AUC (µg.h/ml) | 260 | 858 | 1871–27,103 | 1110 |
| VD (L/kg) | 0.3 | 0.1 | 0.11 | 0.3 |
| Protein binding (%) | 40–54 | 90–93 | 93–98 | 86–90 |
| Terminal half-life (h) | 4–8 | 7–9 | 147–258 | 393 |

Data obtained from Refs [9, 20] and Butterfield et al. Antimicrob Agents Chemother 2011; https://doi.org/10.1128/AAC.01674-10

Abbreviations: *AUC* area under the concentration-time curve, *bid* twice daily, $C_{max}$ peak concentration in serum, *od* once daily, *VD* volume of distribution higher doses of dalbavancin given less frequently produce greater bacterial killing than more frequent smaller doses. However, dalbavancin has slower bactericidal activity than telavancin and oritavancin against *S. aureus* and *Streptococcus pyogenes,* with 24 h required for eradication [18]. Pharmacodynamic studies, however, demonstrated that AUC/MIC best predict its bacterial killing rather than the concentration-based relationship. The $AUC_{24}$/MIC target of 100–300 against *S. aureus* was shown for dalbavancin [19].

### 4.3.7 In Vitro Activity of Dalbavancin

The spectrum of activity of dalbavancin is similar to vancomycin and the other lipoglycopeptide, summarized in Table 4.1. Dalbavancin is more active than vancomycin for *Staphylococcus* species for both methicillin-susceptible and methicillin-resistant strains and similar to telavancin and oritavancin. For MRSA and MSSA strains, the $MIC_{90}$ of dalbavancin is 0.06 μg/ml, while for vancomycin, the $MIC_{90}$ is 1.0 μg/ml, and for *S. epidermidis* and other coagulase-negative staphylococci (CONS) the $MIC_{90}$ of dalbavancin is 0.06–0.12 μg/ml versus 2.0 μg/ml for vancomycin [20]. Dalbavancin is also active against VISA ($MIC_{50}$ and $MIC_{90}$ of 0.25 and 2.0 μg/ml, respectively), but has poor activity against VRSA (MIC >16 μg/ml) [9]. Its activity against streptococci, including *S. pneumoniae,* is also greater than vancomycin and similar to telavancin. Dalbavancin, like telavancin, is active against VRE with the vanB operon but lacks activity against the vanA VRE [9]. VanA results in a modified d-alanyl-d-alanine terminating muropeptide to d-ala-d-lac phenotype.

Dalbavancin has a low potential for resistance development during serial passages of *S. aureus* at sub-MIC concentration, in contrast to vancomycin and teicoplanin which showed four- to eight-fold increase in MIC [4]. However, the very long half-life may predispose to extended exposure of bacteria to subtherapeutic levels that could predispose to resistance. In an in vitro PK/PD model of 28 days after a single dose of 1500 mg dalbavancin against MRSA and MSSA strains, an eight-fold increase in MIC was detected by day 4 in surviving subpopulations of MRSA but increased to 64–128-fold by day 28 [21]. The resistant isolates carried mutations in several different genes, notably walkR, apt, stp 1, and atl.

Clinical experience with dalbavancin has been limited mainly for skin and soft tissue infections, but it has been used for off-label purposes and relative resistance with increased MIC in MRSA infections are now being reported. In one case of central line infection with MRSA initially treated with vancomycin then one dose of dalbavancin to complete the course, grew VISA (blood isolate vancomycin MIC 1.0 μg/ml and urine isolate MIC 4.0 μg/ml) in the urine which was nonsusceptible to dalbavancin (blood isolate MIC 0.015 μg/ml and urine isolate MIC 0.5 μg/ml [22]. The susceptible breakpoint for dalbavancin for EUCAST is ≤0.125 μg/ml and for FDA (2016) is ≤0.25 μg/ml. Genetic alteration in the *yvqF* gene was considered the likely cause of the relative resistance to dalbavancin.

The second case involved a patient with MRSA tricuspid valve endocarditis treated with vancomycin, daptomycin, and dalbavancin after noncompliance with standard daily therapy. After 4 weeks of dalbavancin (total dose 2500 mg), blood cultures grew VISA, resistant to daptomycin and nonsusceptible to dalbavancin (MIC 0.5 μg/ml) [23]. In another case of cardiac device-related endocarditis with MSSA, long-term outpatient therapy with dalbavancin resulted in failure of eradication and isolation of mixed strains of *S. aureus* from blood culture and explanted pacemaker wire [24]. The small colony variants were methicillin resistant, teicoplanin-resistant, and nonsusceptible to dalbavancin (MIC 0.5–1.0 μg/ml). Mutations in *pbp 2* and the DHH domain of GdpP were identified as the most likely explanation.

### *4.3.8 Clinical Efficacy and Safety of Dalbavancin*

Dalbavancin was studied in five phase III RCTs of acute or complicated skin and skin structure infections (SSSIs) with known or suspected gram-positive bacteria. The first international RCT in 2003–2004 enrolled 854 patients and compared IV dalbavancin 1000 mg on day 1 followed by 500 mg on day 8 or IV linezolid 600 mg twice daily and then orally for total of 14 days. *S. aureus* was the most commonly isolated pathogen and 57% were MRSA. Clinical efficacy at test of cure visits was similar, 88.9% for dalbavancin and 91.2% for linezolid, and microbiological eradication were 89.5% and 87.5%, respectively [25]. The second RCT included 565 patients with uncomplicated SSSI to dalbavancin 1000 mg on day 1 with the option of 500 mg on day 8, or IV cefazolin 500 mg every 8 h, with option of switching to oral cephalexin 500 mg every 6 h, conducted in 7 countries. The clinical response was similar between the two groups (89%) [25]. The third phase III RCT (open label) assessed 156 patients with known or suspected MRSA complicated SSSIs randomized to dalbavancin (same dose as the other 2 studies) compared to vancomycin IV 1000 mg every 12 h, with option of switching to oral cephalexin every 6 h for susceptible bacteria. The clinical response was similar for the two groups in evaluable patients, 89.9% for dalbavancin and 86.7% for vancomycin [25].

Two other phase III international double-dummy RCTs were conducted, Discover I and Discover 2, for acute SSSIs suspected or proven gram-positive bacterial infections that require at least 3 days of IV therapy. Dalbavancin 1000 mg was given on day 1 and 500 mg on day 8 compared to vancomycin 1 g or 15 mg.kg every 12 h, with option of switching to linezolid 600 mg every 12 h after day 3 to complete 10–14 days in improved patients for both trials. Discovery I enrolled 573 patients from Europe and North America and Discovery 2 enrolled 729 patients from North America, Europe, Asia, and South Africa [25]. Early clinical response at 48–72 h, time to fever resolution and time to cessation of spread of the infection, and clinical efficacy at day 14 and follow-up at day 28 were similar between dalbavancin and vancomycin/linezolid [25]. Bacteremia in the combined trials was rare, 45 of 1302

(3.4%) patients, and MRSA infections occurred in 162 of 1302 (12.4%) patients, but the response rate did not differ between the groups in patients with bacteremia or MRSA infections.

Safety profile of dalbavancin based on data from phase II and phase III trials showed adverse events were similar or slightly lower than the comparator agents, and the duration of these events did not occur any longer or later [25]. In the combined trials, adverse events occurred in 44.9% on dalbavancin versus 46.8% on comparator agents ($p$ = 0.012), fewer treatment-related adverse events (18.4% vs 20.1%, respectively ($p$ = 0.0014), and fewer serious treatment-related events (0.2 vs 0.7%, respectively, $p$ = 0,021) [26]. The most common adverse events were headache, nausea, diarrhea, vomiting, rash, pruritus, and insomnia. In the Discover trials, the rate of nephrotoxicity of dalbavancin two doses ($n$ = 637) compared to at least 10 days on vancomycin ($n$ = 54) was fewer 3.3% vs 9.3%, respectively, $p$ = 0.06 [26].

### 4.3.9 Off-Label Use of Dalbavancin

The long half-life of dalbavancin and its potent activity against staphylococci, streptococci, and enterococci with once a week parenteral dosing is appealing for outpatient therapy especially in people who use drugs (PWUD) and for infections requiring long-term therapy such as infective endocarditis and osteomyelitis. In recent years, there have been several reports of dalbavancin use in these settings. In a retrospective study of 56 patients (30% PWUD), dalbavancin (71%) or oritavancin (25%) or both (4%) were used to complete treatment, after previous antibiotics for 7–24.5 days (median 13 days) for abscess with SSSIs (36%), osteomyelitis (27%), and endocarditis (9%) [27]. The most common pathogens were MSSA (25%), MRSA (19%), *E. faecalis*, and CoNS (11%). Clinical failure occurred in 15% of cases and 18% were lost to follow-up. It was estimated that the total reduction in hospital length of stay was 514 days and cost-saving was $963,456 [27].

In another retrospective case series of 32 patients in PWUD, dalbavancin was used as a secondary agent to complete treatment on discharge where no acceptable oral agent was available for serious *S. aureus* infections (endocarditis, osteomyelitis, septic phlebitis, epidural infection) [28]. Most of the infections were due to MRSA (88%), and vancomycin was the most common previously used antibiotic (average of 13 days). Ten (31%) were lost to follow-up and 4 (13%) failed therapy and the majority who completed treatment had clinical response [28]. A report from Baltimore, however, is of concern with 33% failure of therapy with dalbavancin for gram-positive bacteremia and endocarditis where standard outpatient parenteral therapy could not be arranged (50% PWUD) [29]. This is in contrast to the experience in Vienna where dalbavancin was used as primary and sequential therapy for gram-positive bacteremia with endocarditis resulting in clinical success in 92.6% [30]. In 24 of 27 patients, dalbavancin was used only after clearance of the bacteremia from the blood stream. Over 90% of the bacteria were susceptible to β-lactam agents, five patients had prosthetic valves, and another five had cardiac device-related endocarditis. Ten patients (37.0%) received once weekly regimen and 17 (63.0%) received twice weekly regimen.

The Spanish experience was retrospectively analyzed for clinical efficacy and cost-effectiveness of dalbavancin as consolidation therapy in gram-positive bacterial endocarditis and blood stream infections from 14 hospitals [31]. Eighty-three patients (mean age 73 years) were included of which 59.1% were blood stream infections (41% complicated) and 40.9% were infective endocarditis (44.1% with prosthetic valves). Microorganisms recovered included CoNS (44.1%), *S. aureus* (29.4%, 11.8% MRSA), 11.8% streptococci, and 8,8% *E. faecalis.* The clinical effectiveness of dalbavancin in treatment of endocarditis was 96.7% and for bacteremia 100%. The saving in hospital stay was 636 days for bacteremia (€315,424.20) and 557 days for infective endocarditis (€283,187.45) [31].

A phase II, single-center, open-label RCT was conducted in a tertiary care teaching hospital in Ukraine to assess dalbavancin for the treatment of osteomyelitis [32]. Eighty patients were randomized to receive dalbavancin ($n = 70$) or standard of care ($n = 10$) for 4–6 weeks. *S. aureus* (60%) was the most common pathogen (32/42 MSSA) and others included streptococci, CoNS, and enterococci. Clinical efficacy at 42 days was 97% in the dalbavancin group and 88% in the standard treatment group. The clinical efficacy was maintained in the test group at 1 year (96%). The standard treatments were vancomycin IV or vancomycin for 5–16 days and then switch to IV linezolid or IV levofloxacin to complete 29 days [32]. Adverse events occurred in 10 (14.3%) patients on dalbavancin but only 1 was considered drug related. Larger multicenter RCTs are needed to establish its role in the treatment of osteomyelitis.

### 4.3.10 Oritavancin

Oritavancin (Orbactiv) was approved in 2014 in the US for treatment of acute bacterial SSSIs by gram-positive pathogens. It is a synthetic derivative of a naturally occurring glycopeptide (chloroeremomycin) by adding of N-alkyl-p-chlorophenylbenzyl to the disaccharide sugar [9]. This alteration improves the activity against vancomycin-susceptible enterococci and VRE, including VanA strains. The lipophilic side chain anchors the drug to the bacterial cell membrane and improves the binding to its target, including D-Ala-D-Lac found in VanA enterococci [9]. Similar to the other lipoglycopeptide, oritavancin inhibits cell wall synthesis and increases membrane permeability, and it can disrupt membrane potential in both stationary and the exponential growing phase of bacteria. Thus, in vitro it can kill stationary phase and biofilm of *S. aureus* [33], which is unusual.

### 4.3.11 Pharmacokinetics and Pharmacodynamics of Oritavancin

Oritavancin, similar to the other lipoglycopeptides, has high protein binding (86–90%) and a terminal half-life of 393 h (see Table 4.2) which allows for once weekly dosing [9] or potentially once every 2 weeks. It is not metabolized by the liver, and it is excreted slowly in the urine and feces over 14 days. It accumulates extensively in cells with high retention and slow clearance from the liver, spleen, kidneys, and lungs with a volume of distribution of 1 L/kg. Oritavancin concentration in blister fluid is 19% of the plasma levels but the concentration in alveolar macrophages is 142-fold the serum level [20]. No dose adjustment is required for renal or hepatic impairment. Studies indicate that oritavancin can be a weak inhibitor or inducer of cytochrome P450 enzymes and may inhibit the metabolism of warfarin [34].

Oritavancin has rapid bactericidal activity in general but slower against VRE and high inoculum of VISA and demonstrates concentration-dependent microbial killing activity [20]. For the three new glycopeptides, the PK/PD parameters best predictive of efficacy are free AUC/MIC and the $C_{max}$/MIC ratio. In a phase II study of oritavancin in patients with *S. aureus* bacteremia, it was demonstrated that $AUC_{0-24}$/MIC, $C_{max}$, and fT > MIC (free drug % time > MIC) all had significant correlation with clinical efficacy [35]. Reduction of bacterial load of 1–1.5 cfu was achieved with free drug $C_{max}$/MIC ratio of about 14 and *f*T > MIC of 42–50%. The free drug % time > MIC was the PK/PD parameter most strongly associated with efficacy [35]. The drug can be administered as a single IV infusion of 1200 mg over 3 h in adults and repeat dosing is not required for treatment acute bacterial SSSIs. This dose in population pharmacokinetic studies results in $C_{max}$ of 138 µg/ml (20.7 µg/ml free fraction) and the AUC of 1110 µg.h/ml (165 µg/ml free fraction) for *f* $C_{max}$/MIC and *f*$AUC_{24}$/MIC of 165 and 1352 h, respectively [20]. In vitro a $C_{max}$/MIC ratio of 4 of oritavancin can result in a 3-log killing of an inoculum ($10^5$cfu) of MRSA [34].

### 4.3.12 In Vitro Activity of Oritavancin

The spectrum of activity of oritavancin is similar to vancomycin against vancomycin-susceptible strains of staphylococci, streptococci, and enterococci; the three lipoglycopeptides are similar in activity, but more potent than vancomycin (see Table 4.1). Oritavancin is slightly more potent than the other compounds against enterococci and it is effective against VRSA and VRE, including strains with *vanA* genotype [20]. Its antimicrobial activity is reduced against vancomycin intermediate or resistant strains with $MIC_{90}$ of 1 µg/ml and 2 µg/ml for VRSA and VISA, respectively [34]. All three lipoglycopeptides are active against a broad range of aerobic and anaerobic gram-positive bacteria (cocci and bacilli). The $MIC_{90}$ for oritavancin against *Micrococcus* species, *Listeria monocytogenes*, and *Corynebacterium* species is <0.06 μg/ml; for *C. difficile* and *C. perfringens* 1.0 μg/ml; for *Peptococcus* and *Peptostreptococcus* species 0.5 μg/ml; and for *Propionibacterium* species 0.25 μg/ml [34]. Overall, >99% of *S. aureus* (MSSA and MRSA), *Streptococcus* spp., and *Enterococcus* spp. are susceptible to oritavancin with in vitro activity equal to or greater than vancomycin, daptomycin, linezolid, or tigecycline [36].

Oritavancin and telavancin are effective in vitro and in animal model (telavancin) in killing MSSA, MRSA, and VISA in biofilms [20]. Also oritavancin cellular-to-extracellular concentration ratio is the highest in macrophages compared to gentamicin, azithromycin, telithromycin, ciprofloxacin, and moxifloxacin [37]. Moreover, acidic pH 5 (as found in phagolysosomes) reduces the activity of the aminoglycoside, macrolides, and the quinolones but not oritavancin [37]. Oritavancin produces a post-antibiotic effect in vitro against MRSA and VRE with concentration-dependent duration and shorter for staphylococci than for enterococci [9]. However, the significance of the post-antibiotic effect on clinical response is unclear.

The susceptibility breakpoint criteria (package insert) for oritavancin for *S. aureus* and *E. faecalis* is ≤0.12 μg/ml and ≤ 0.25 μg/ml for *Streptococci* species, based on broth microdilution methods. But the broth test medium should be supplemented with polysorbate 80 (0.002% concentration) to prevent adsorption of the compound to the plastic wells.

In vitro moderate resistance to oritavancin can occur in enterococci strains with the genes of vanA and vanB, but resistance in strains with *vanB* operon was mainly observed when induced by teicoplanin or constitutively expressed [38]. In isolates with *vanA*, gene cluster expression of *vanZ* may also develop resistance to oritavancin with MIC of 8 μg/ml [38]. High level resistance to oritavancin has been difficult to produce in the laboratory or found in clinical pathogens to date [9, 36]. One study reported it was possible to induce resistance to oritavancin in *Enterococci* strains with serial passaging over 20 days with elevated MIC up to 32-fold higher than baseline [39]. Interestingly not all the isolates showed cross-resistance to the other lipoglycopeptides suggesting different modes of resistance exist among the group. However, high level resistance was not induced to oritavancin as the highest MIC (even with 32-fold increase) was 1 μg/ml, whereas even with a non-vanA isolate the MIC increased >16 μg/ml with dalbavancin and 8 μg/ml with telavancin.

### 4.3.13 *Clinical Efficacy of Oritavancin*

The single dose of 1200-mg oritavancin for treatment of acute bacterial SSSIs was derived from the phase 2 dose-ranging study, SIMPLIFI [40]. Two phase 3 double-blinded, multicenter, noninferiority RCTs of acute bacterial SSSIs, SOLO 1 and 2, enrolled 1959 patients and led to the FDA approval [34]. These studies included wound infections, cellulitis, and large skin abscesses that were considered by the clinicians to require 7 days of parenteral therapy. A single IV dose of 1200 mg oritavancin was compared to IV vancomycin 1 g or 15 mg/kg every 12 h for 7–10 days.

The clinical responses in the two studies were similar between the two treatments and met the noninferiority guideline. MRSA accounted for 42.8% of the 945 *S. aureus* isolated and the clinical response was similar for the 2 regimens in patients infected with these pathogens, clinical cure 83.3% for oritavancin and 84.1% for vancomycin [34].

Drug-related adverse events and severe adverse events were similar between oritavancin (22.8% and 5.2%) and vancomycin (28.4% and 4.9%), but increase in alanine aminotransferase (ALT) was slightly higher in patients on oritavancin (2.3–3.2% vs 1.0–2.0%) [34]. Nausea was the most common side effects to oritavancin (9.9%) with similar rate as vancomycin (10.5%).

### 4.3.14 Off-Label Use of Oritavancin

There are several reports of off-label use of oritavancin, mainly small retrospective studies. In one study, 17 patients were treated with multiple doses or oritavancin for osteomyelitis, intravascular infections, pneumonia, and complicated surgical site infections with clinical success in all patients [41]. Four patients (24%) had adverse event requiring discontinuation of therapy (two infusion related) with improvement. Most of the patients received prior therapy and were switched to oritavancin to facilitate outpatient therapy, and, thus, efficacy cannot be accurately assessed. In another report of the use of oritavancin for continued therapy after prior antibiotic treatment of 10 patients with various gram-positive infections, similar limitations preclude proper assessment for efficacy [42]. However, one patient treated with 3 doses of oritavancin and previous course of vancomycin experienced hearing loss. Thus, patients receiving sequential therapy with these two drugs should be carefully assessed for auditory disturbance.

Osteomyelitis is one of the infections where once weekly IV therapy with oritavancin would be suitable as staphylococci are the most common pathogens, and it requires prolonged (4–6 weeks) therapy, commonly administered parenterally. A review of oritavancin treatment of osteomyelitis from case series and the CHROME registry was recently published [43]. Twenty-three cases were included in the review with MRSA in 48%, the most common pathogen. Although clinical success was present in 87%, it may not be primarily due to oritavancin as 14 patients (56%) had prior antibiotics and 5 cases received concomitant antimicrobial agents which could have affected the outcome. The best data on treatment of acute osteomyelitis with oritavancin was also recently published. This study was a 2 year, multicenter, retrospective, descriptive study of 134 patients with predominantly MRSA (71.9%) osteomyelitis with no concurrent antibiotics, but 18 (13.4%) received previous therapy [44]. Patients received oritavancin 1200 mg IV infusion followed by 800 mg once weekly for 4–5 weeks. Of the 130 patients completing post-therapy evaluation at 3 and 6 months, 104 (80.0%) achieved clinical success. Adverse events were reported in only five (3.7%) patients, hypoglycemia-related symptoms ($n$ = 3), tachycardia during infusion ($n$ = 2), and chest pain in one. The infusion was discontinued in the latter 2 patients and symptoms resolved after treatment with diphenylhydramine. Both patients eventually completed the course of oritavancin [44].

Bacterial endocarditis is another area where oritavancin may have potential benefit over standard therapy such as vancomycin for MRSA or enterococci infection. In the rabbit model of aortic valve endocarditis with MRSA, oritavancin was found to be as effective as vancomycin in reducing bacterial counts on vegetations [45]. It was also shown to significantly reduce the bacterial counts in vegetations infected with VRE in a similar model, enhanced by addition of gentamicin which prevented emergence of resistance [46]. However, there is very little human data on treatment of infective endocarditis with oritavancin. In a review of gram-positive bacterial infections treated with oritavancin at a single medical center, 4 of 75 patients had endocarditis and all were cured, but details including the etiologic pathogens were not provided [47]. A single case of prosthetic valve endocarditis with VRE and recurrent bacteremia on daptomycin and tigecycline was cured after 7 weeks of oritavancin, but its role is difficult to assess since 4 weeks of linezolid and tigecycline were given before with negative blood cultures [48].

### *4.3.15 Comments*

Although the three lipoglycopeptides are more potent than vancomycin and are effective against vancomycin-resistant staphylococci and enterococci, especially oritavancin, clinical trials have not shown superior efficacy. However, vancomycin-nonsusceptible gram-positive pathogens were rare or absent in these trials. Oritavancin and dalbavancin have the advantage of single-dose therapy for complicated skin and soft tissue infections which offers convenience and potential for cost-savings from reduced days in hospital. This is especially relevant in the treatment PWUDs. However, potentially their long half-life could be a disadvantage in the case of drug toxicity with prolonged effect, but this was not demonstrated in any of the trials.

Telavancin was approved for HAP including VAP due to MRSA but it had no clear benefit over vancomycin and slightly higher occurrence of Q-Tc-interval prolongation. Both oritavancin and telavancin should have advantage over vancomycin for prosthetic gram-positive infections and endocarditis in view of their rapid bactericidal activity and killing effect on biofilms. Hence, future multicenter trials are warranted especially for MRSA infections compared to vancomycin. Oritavancin and/or dalbavancin should be compared to vancomycin for treatment of osteomyelitis due to MRSA or MSSA in patients with severe β-lactam allergy, as they offer convenience and decrease healthcare cost. However, it is now evident from controlled randomized trials that the common practice of long-term parenteral therapy for bacterial endocarditis and osteomyelitis is not necessary in most cases, as short-term IV therapy followed by oral agents with very good bioavailability is just as effective [48]. Thus, cost-effective studies comparing the long-acting glycopeptides given parenterally at 1–2 week intervals versus standard short-term IV therapy followed by oral agents would be appealing.

Comparative studies with the new glycopeptides, especially oritavancin, should be done in patients with VRE bacteremia. There is limited therapeutic options for this condition, most commonly used are daptomycin and linezolid, and the mortality rates with current treatment are high (20–46%); thus more effective drugs for therapy are needed.

Despite the potential value of the new glycopeptides, they should be used selectively for special circumstances as widespread overuse will lead to development of progressive resistant mutations.

## References

1. Blaskovich MAT, Hansford KA, Butler MS, Jia Z-G, Mark AE, Cooper MA. Developments in glycopeptide antibiotics. ACS Infect Dis. 2018;4:715–35.
2. Binda E, Marinella F, Marcone GL. Old and new glycopeptide antibiotics: action and resistance. Antibiotics. 2014;3:572–94.
3. Marcone GL, Binda E, Berini F, Martinelli F. Old and new glycopeptides: from product to gene and back in the post-genomic era. Biotechnol Adv. 2018;36:534–54.
4. Butler MS, Hansford KA, Blaskovich MA, Halai R, Cooper MA. Glycopeptide antibiotics: back to the future. J Antibiot. 2014;67:631–44.
5. Higgins DL, Chang R, Debalov DV, et al. Telavancin, a multifunctional lipoglycopeptide, disrupts both cell wall synthesis and cell membrane integrity in methicillin-resistant *Staphylococcus aureus*. Antimicrob Agents Chemother. 2005;49:1127–34.
6. Charmeski L, Patel PN, Sym D. Telavancin: a novel lipoglycopeptide antibiotic. Ann Pharmacother. 2009;43:928–37.
7. Wenzler E, Rodvold KA. Telavancin: the long winding road from discovery to Food and Drug Administration approvals and future directions. Clin Infect Dis. 2015;61(suppl. 2):S38–47.
8. Hegde SS, Reyes N, Wiens T, et al. Pharmacodynamics of telavancin [TD-6424], a novel bactericidal agent, against gram-positive bacteria. Antimicrob Agents Chemother. 2004;48:3043–50.
9. Zhanel GC, Calic D, Schweizer F, et al. New lipoglycopeptides. A comparative review of dalbavancin, oritavancin and telavancin. Drugs. 2010;70:859–86.
10. Rybak MJ, Le J, Lodise TP, et al. Therapeutic monitoring of vancomycin for serious methicillin-resistant *Staphylococcus aureus* infections: a revised consensus guideline and review of the American Society of Health-System Pharmacists, the Infectious Disease Society of America, the Pediatric Infectious Disease Society, and the Society of Infectious Diseases Pharmacists. Am J Health Syst Pharm. 2020;77:835–64.
11. Atwood RJ, LaPlante KL. Telavancin: a novel lipoglycopeptide. Antimicrob Agent Am J Health Syst Pharm. 2007;64:2335–48.
12. Swartz TH, Huprikar S, Lumbardi V, Pinney S, Anyanwu A, Lee M, Patel G. Heart transplant in a patient with heteroresistant vancomycin-intermediate *Staphylococcus aureus* ventricular assist device mediastinitis and bacteremia. Transpl Infect Dis. 2013;15:E177–81.
13. Stryjewski ME, O'Riordan WD, Lau WK, et al. Telavancin versus standard therapy for treatment of complicated skin and soft tissue infections due to gram-positive bacteria. Clin Infect Dis. 2005;40:1601–7.
14. Stryjewski ME, Chu VH, O'Riordan WD, et al. Telavancin versus standard therapy for treatment of complicated skin and skin structure infections caused by gram-positive bacteria: FAST 2 study. Antimicrob Agents Chemother. 2006;50:862–7.

15. Stryjewski ME, Graham DR, Wilson SE, et al. Telavancin versus vancomycin for complicated skin and skin-structure infections caused by gram-positive organisms. Clin Infect Dis. 2008;46:1683–93.
16. Rubenstein E, Lalani T, Corey GR, et al. Telavancin versus vancomycin for hospital-acquired pneumonia due to gram-positive pathogens. Clin Infect Dis. 2011;52:31–40.
17. Luethner K, Buechler K, Kogan D, Saguros A, Lee S. Clinical efficacy of dalbavancin for the treatment of acute bacterial skin and skin structure infections [ABSSSI]. Ther Clin Risk Manag. 2016;12:931–40.
18. Goldstein BP, Draghi DC, Sheehan DJ, et al. Bactericidal activity and resistance development profiling of dalbavancin. Antimicrob Agents Chemother. 2007;51:1150–4.
19. Andes D, Craig WA. In vivo pharmacodynamics activity of the glycopeptide dalbavancin. Antimicrob Agents Chemother. 2007;51:1633–42.
20. Van Bambeke F. Lipoglycopeptide antibacterial agents in gram-positive infections: a comparative review. Drugs. 2015;75:2073–95.
21. Werth BJ, Ashford NK, Penewit K, et al. Dalbavancin exposure in vitro selects for dalbavancin-non-susceptible and vancomycin-intermediate strains of methicillin-resistant *Staphylococcus aureus*. Clin Microbiol Infect. 2020:S1198-743x[20]30508. https://doi.org/10.1016/jmi.2020.08.025.
22. Werth BJ, Jain R, Hahn A, et al. Emergence of dalbavancin non-susceptible, vancomycin-intermediate *Staphylococcus aureus* [VISA] after treatment of MRSA central line-associated bloodstream infection with a dalbavancin- and vancomycin-containing regimen. Clin Microbiol Infect. 2018;24(4):429e1–5. https://doi.org/10.1016/jcmi.2017.07.028.
23. Steele JM, Seabury RW, Hale CM, Mogale BT. Unsuccessful treatment of methicillin-resistant *Staphylococcus aureus* endocarditis with dalbavancin. Clin Pharm Ther. 2018;43:101–3.
24. Kussmann M, Karer M, Obermueller M, et al. Emergence of a dalbavancin induced glycopeptide/lipoglycopeptide non-susceptible *Staphylococcus aureus* during treatment of a cardiac devise-related endocarditis. Emerg Microbes Infect. 2028;7:202. https://doi.org/10.1038/s41426-018-0205-z.
25. Vardakas KZ, Falagas ME. Dalbavancin: re-enforcing the arsenal against gram-positive bacteria causing skin and skin structure infections. Clin Invest. 2014;4:63–75.
26. Dunne MW, Talbot GH, Boucher HW, Wilcox M, Puttagunta S. Safety of dalbavancin in the treatment of skin and skin structure infections: a pooled analysis of randomized, comparative studies. Drug Saf. 2016;39:147–57.
27. Morrisette T, Miller MA, Montague BT, Barber GR, McQueen RB, Krsak M. On-and off-label utilization of dalbavancin and oritavancin for gram-positive infections. J Antimicrob Chemother. 2019;74:2405–16.
28. Bryson-Cahn C, Beieler AM, Chan JD, Harrington RD, Dhanireddy S. Dalbavancin as secondary therapy for serious *Staphylococcus aureus* infections in a vulnerable patient population. Open forum. Infect Dis. 2019;6:ofz028. https://doi.org/10.1093/ofid/ofz028.
29. Ajaka L, Heil E, Schmalzle S. Dalbavancin in the treatment of bacteremia and endocarditis in people with barriers to standard care. Antibiotics. 2020;9:0700.
30. Tobudic S, Forstner C, Burgmann H, et al. Dalbavancin as primary and sequential treatment for gram-positive infective endocarditis: 2-year experience at the general hospital of Vienna. Clin Infect Dis. 2018;67:795–8.
31. Hidalgo-Tenorio C, Vinuesa D, Plata A, et al. DABACEN cohort: dalbavancin as consolidation therapy in patients with endocarditis and/or bloodstream infection produced by gram-positive cocci. Ann Clin Microbiol Antimicrob. 2019;18:30. https://doi.org/10.1186/s12941-019-0329-6.
32. Rappo U, Puttagunta S, Shevchenko V, et al. Dalbancin for the treatment of osteomyelitis in adult patients: a randomized clinical trial of efficacy and safety. Open forum. Infect Dis. 2019;6:ofy331. https://doi.org/10.1093/ofid/ofy331.
33. Belley A, Neesham-Grenon E, McKay G, et al. Oritavancin kills stationary-phase and biofilm *Staphylococcus aureus* cells in vitro. Antimicrob Agents Chemother. 2009;53:918–25.

34. Stein GE. Oritavancin: a long-half-life lipoglycopeptide. Clin Infect Dis. 2015;61:727–32.
35. Bhavnani SM, Passarell JA, Owen JS, Loutit JS, Porter SB, Ambrose PG. Pharmacokinetic-pharmacodynamic relationships describing the efficacy of oritavancin in patients with *Staphylococcus aureus* bacteremia. Antimicrob Agents Chemother. 2006;50:994–1000.
36. Fluh G, Seifert H, Kaasch AJ. Oritavancin: an update. Future Microbiol. 2018;13:727–9.
37. Seral C, Van Bambeke F, Tulkens PM. Quantitative analysis of gentamicin, azithromycin, telithromycin, ciprofloxacin, moxifloxacin, and oritavancin [LY333328] activities against intracellular *Staphylococcus aureus* in mouse J774 macrophages. Antimicrob Agents Chemother. 2003;47:2283–92.
38. Arthur M, Depardieu F, Reynolds P, et al. Moderate-level resistance to glycopeptide LY333328 mediated by genes of vanA and vanB clusters in enterococci. Antimicrob Agents Chemother. 1999;43:1875–80.
39. Arhin FF, Seguin DL, Belley A, Moeck G. *In vitro* stepwise selection of reduced susceptibility in lipoglycopeptides in enterococci. Diag Microbiol Infect Dis. 2017;89:168–71.
40. Dunbar LM, Milata J, McClure TY, et al. Comparison of the efficacy and safety of oritavancin front-loaded dosing regimens to daily dosing: an analysis of the SIMPLIFI trial. Antimicrob Agents Chemother. 2011;55:3476–84.
41. Schulz LT, Dworkin E, Del-Pena J, Rose WE. Multiple-dose oritavancin in a retrospective cohort of patients with complicated infections. Pharmacotherapy. 2018;38:152–9.
42. Stewart CL, Turner MS, Frens JJ, Snider CB, Smith JR. Real-world experience with oritavancin therapy in invasive gram-positive infections. Infect Dis Ther. 2017;6:277–89.
43. Scoble PJ, Reilly J, Tillotson GS. Real-world oritavancin for the treatment of osteomyelitis. Drugs Real World Outcomes. 2020;7(Suppl 1):546–54.
44. Van Hise NW, Chundi V, Didwania D, et al. Treatment of acute osteomyelitis with once weekly oritavancin: a two-year, multicenter, retrospective study. Drugs Real World Outcomes. 2020;7(Suppl 1):41–5.
45. Kaatz GW, Seo SM, Aeschilmann JR, et al. Efficacy of LY333328 against experimental methicillin-resistant *Staphylococcus aureus* endocarditis. Antimicrob Agents Chemother. 2006;50:981–3.
46. Saleh-Mghir A, Lefort A, Ptegnief Y, et al. Activity and diffusion of LY333328 in experimental endocarditis due to vancomycin-resistant *enterococcus faecalis*. J Antimicrob Agents Chemother. 1999;43:115–20.
47. Brownell LE, Adamsick M, McCleary EK, Vanderloo JP, Ernst EJ, Jackson ER, Schulz LT. Clinical outcomes and economic impact of oritavancin for gram-positive infections: a single academic medical center experience. Drugs Real World Outcomes. 2020;7(Suppl 1):13–9.
48. Fong IW. Issues in therapeutics of some bacterial infections: vancomycin use, osteomyelitis, endocarditis, and *Staphylococcus aureus* bacteremia. In: Current trends and concerns in infectious diseases. Cham: Springer Nature; 2020.

# Chapter 5
# A New Fluoroquinolone: Delafloxacin

## 5.1 Introduction

The fluoroquinolones (FQs) have been in clinical use for over 50 years for a wide spectrum of clinical infections and could be used both intravenously and orally due to their good oral bioavailability. Besides community-acquired pneumonia, they were used mainly for gram-negative bacilli infections from a variety of sources. However, with widespread and indiscriminate use, their efficacy and reliability have been eroded by progressive global resistance. Moreover, over the years there has been several black box warnings for the FQ class besides high risk for developing *Clostridium difficile* colitis. These include tendinitis, tendon rupture (Achilles tendon), peripheral neuropathy, aseptic meningitis, myasthenia gravis exacerbation, significant decrease in blood glucose and certain mental effects, QTc prolongation, and aortic aneurysm or dissection in patients with atherosclerotic disease [US Food and Drug Administration, Safety Announcement, 12-20-2018]. The United States (US) Food and Drug Administration (FDA) in 2016 issued a warning that the adverse effects associated with the FQ may outweigh the benefits in certain infections.

The question may, thus, be asked: do we need another FQ antibiotic? Since 2015 the World Health Organization (WHO) and health experts in the US and Europe have been advocating for development of new antimicrobials to combat the ever-increasing global antibiotic resistance. Hence, the development of a new broader-spectrum FQ is one of the response for "call-to-arms" for new antimicrobials to fight the global pandemic of antimicrobial resistance.



I. W. Fong, *New Antimicrobials: For the Present and the Future*, Emerging Infectious Diseases of the 21st Century,
https://doi.org/10.1007/978-3-031-26078-0_5

## 5.2 Pharmacology and Pharmacokinetics

Delafloxacin is a chemically unique fully synthetic anionic FQ approved in several countries to treat acute bacterial skin and skin structure infections (ABSSSI), including the US and the European Union (EU), but also community-acquired pneumonia (CAP) in the US. It is marketed as Quofenix™ in the EU and Baxdela® in the US. Modifications of the FQ structure were made to enhance its antibacterial spectrum and activity and improve the pharmacokinetic and toxicity profiles. The three changes of its chemical structure distinguishing it from the other FQ include (i) aromatic substitution at N1 position to enhance the antibacterial activity; (ii) addition of chlorine at C8 position to improve its activity against FQ-resistant gram-positive bacteria; and (iii) the lack of a strongly basic group at C7 position (see Fig. 5.1) which renders it weakly acidic which enhances its antibacterial potency in acidic environment such as abscesses [1]. Delafloxacin mainly exists as an uncharged molecule at low pH, allowing easy transmembrane transition to concentrate in the bacterium, and once in the cytoplasm with neutral pH, it becomes anionic, whereas other FQ remains in the zwitterion form in the bacterium [2].

The oral absorption of delafloxacin is rapid, but the mean bioavailability of 58.8% is lower than other FQs which are nearly 100% [3]. Hence, the oral dose of 450 mg will provide equivalent serum levels and area under the concentration curve (AUC) as 300 mg intravenously (IV). Following oral dosing, the peak serum concentration occurs at 1–2.5 h in volunteers and at the end of an hour infusion after IV dosing [1]. The peak serum concentration ($C_{max}$) and $AUC_{0-12}$ ranged from 8.9 to 9.3 μg/mL and 21.8 to 23.4 μ.h/mL, respectively, after 300 mg IV every 12 h, and 6.1 μg/mL and 24.2 μg.h/mL following 450 mg oral dose [3, 4]. The protein binding

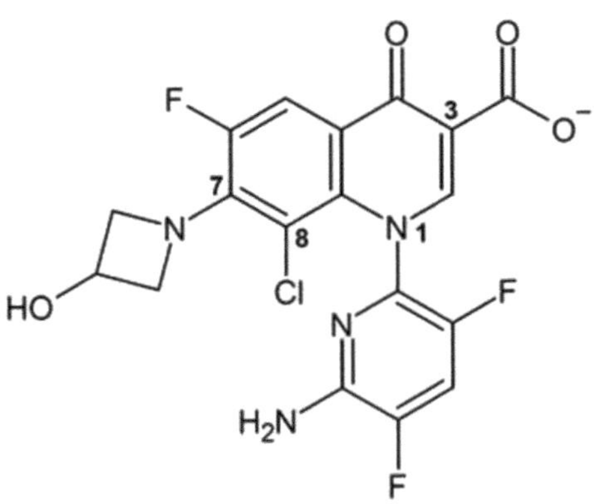


**Fig. 5.1** Chemical structure of delafloxacin and numbering of atoms for the key positions
Note: The changes to the chemical structure distinguishing delafloxacin from other fluoroquinolones (FQ) are at position (i) aromatic substitution at N1 enhances the antibacterial activity; (ii) addition of chlorine at position C8 improve its activity against FQ-resistant bacteria; (iii) lack of strong basic group at position C7 enhances its antibacterial potency in acidic environment (weakly acidic)

of delafloxacin is about 84%, and the volume of distribution is equivalent to the total body water, 30.2–38.5 L [1]. Following IV or oral administration, delafloxacin demonstrates a biexponential decay in plasma concentrations with half-life of about 6–8 h and steady state is achieved in 3 days after multiple dosing. Renal elimination accounts for 40% of the total body clearance, 50% excreted unchanged in the urine, and the primary metabolic pathway is through glucuronidation. As with other FQs, absorption of oral delafloxacin is significantly reduced with antacids containing aluminum or magnesium, sucralfate, iron, zinc, or formulations containing divalent and trivalent cations and should be taken ≥2 h before or 6 h after [2].

No dosing adjustment is needed for hepatic or mild-moderate renal impairment, but in severe renal impairment (creatinine clearance <30 mL/min), the IV dose should be 200 mg every 12 h, but no dosage adjustment is needed for the oral dose of 450 mg every 12 h [1]. No significant drug-drug interactions are recognized with delafloxacin.

## 5.3 Pharmacodynamics

Delafloxacin, similar to other FQs, demonstrates concentration-dependent bactericidal activity, and the efficacy is best predicted by the pharmacodynamic indices of maximal concentration of free drug ($fC_{max}$)/MIC (minimal inhibitory concentration) or the area under the concentration curve (AUC)/MIC ratio of 100–125 against gram-negative bacteria, and AUC/MIC ratio of 30–50 against gram-positive pathogens [1].

## 5.4 In Vitro Activity

Unlike other FQs, delafloxacin has dual (nearly equivalent) affinity for the essential bacterial enzymes DNA gyrase and topoisomerase IV; DNA gyrase is more susceptible to inhibition in gram-negative bacteria and the topoisomerase IV is more susceptible to inhibition in gram-positive pathogens [5, 6]. While other FQs (ciprofloxacin, moxifloxacin) have reduced activity in acidic environment, delafloxacin antibacterial activity is enhanced at low pH which is present in abscesses, phagolysosomes, and inflammatory cells [7]. It also has good activity against *Staphylococcus aureus* (methicillin-sensitive [MSSA] and methicillin-resistant [MRSA]) mature biofilms at clinically achievable concentrations by reducing viability >50% in 24–48 h [8].

Delafloxacin has a very broad spectrum of activity against gram-positive and gram-negative bacteria. The US Food and Drug Administration (FDA) susceptibility breakpoints for delafloxacin is shown in Table 5.1 [9]. The in vitro activities of delafloxacin has been compared to a large number of comparative agents for gram-positive and gram-negative pathogens collected in Europe and the US in 2014 (6485

**Table 5.1** FDA susceptibility breakpoints for broth microdilution method

| Pathogen | Susceptible | Intermediate | Resistance (μg/mL) |
|---|---|---|---|
| *S. aureus* (MSSA/MRSA) | ≤0.25 | 0.5 | ≥1.0 |
| *S. haemolyticus* | ≤0.25 | 0.5 | ≥1.0 |
| *S. pyogenes* | ≤0.06 | – | – |
| *S. agalactiae* | ≤0.06 | 0.12 | >0.25 |
| *S. anginosus* gp. | ≤0.06 | – | – |
| *E. faecalis* | ≤0.12 | 0.25 | ≥0.5 |
| *Enterobacteriaceae* | ≤0.25 | 0.5 | ≥1.0 |
| *P. aeruginosa* | ≤0.5 | 1.0 | ≥2.0 |

Data obtained from ref. 9
Abbreviations: *MSSA* methicillin-susceptible *S. aureus*, *MRSA* methicillin-resistant *S. aureus*, – no criteria

isolates) and between 2014 and 2016 (36,383 isolates); see Tables 5.2 and 5.3 [10]. It had the lowest $MIC_{90}$ compared to other agents against MSSA, MRSA, *S. pneumoniae, S. viridans* group, and β-hemolytic *Streptococci* and similar activity to ciprofloxacin and levofloxacin to the *Enterobacteriaceae* [10]. The susceptibility rates of *Pseudomonas aeruginosa* to delafloxacin have been found to be similar to ciprofloxacin, 74.0% and 75.0%, respectively [1]. Delafloxacin has potent activity (similar to or greater than other FQs) against *Haemophilus influenzae*, *Moraxella catarrhalis*, *Neisseria gonorrhoeae*, *Legionella* spp., *Chlamydia pneumoniae*, *Helicobacter pylori*, *Mycobacterium tuberculosis*, and *Mycobacterium avium* complex [1, 10]. Compared to ciprofloxacin and moxifloxacin in a susceptibility study of 131 isolates of nontuberculous mycobacteria, delafloxacin was the most active against *Mycobacterium fortuitum* and *M. mucogenicum* groups and *M. kansasii* with $MIC_{50}$ of 0.12 to 0.5 μg/mL [11].

### 5.4.1 Microbial Resistance

Resistance to the FQs occurs by mutations in the genes controlling the target enzymes and by drug efflux pumps. Mutations in the genes (QRDR) encoding subunits of DNA gyrase (*gyr*A and *gyr*B) and topoisomerase IV (*gr*IA and *gr*IB in gram-positive species; *par*C and *par*E in gram-negative species) result in alterations in the target enzyme configuration and decrease FQ binding that leads to resistance [1]. Compared to other FQs, delafloxacin has greater stability to target enzyme mutations in gram-positive bacteria and whereas single point mutations at both *gr*A and *gr*IA in MRSA produced resistance to ciprofloxacin, levofloxacin, and moxifloxacin, this did not result in delafloxacin resistance [1]. Resistance of MRSA isolates to delafloxacin required triple mutations in the QRDR. Compared to other FQs, delafloxacin had lower mutation prevention concentration in respiratory pathogens (*S. aureus*, *S. pneumonia*, *H. influenzae*, and *M. catarrhalis*) [12].

**Table 5.2** Comparative in vitro activities of delafloxacin and comparators against gram-positive pathogens

| Organisms (no.)/antibiotic | % Susceptible | MIC-50 (μg/mL) | MIC-90 (μg/mL) |
|---|---|---|---|
| **MSSA (777)** | | | |
| Delafloxacin | – | ≤0.004 | 0.008 |
| Levofloxacin | 89.8 | 0.25 | 2.0 |
| Ceftaroline | 100 | 0.25 | 0.25 |
| Clindamycin | 94 | ≤0.25 | ≤0.25 |
| Oxacillin | 100 | 0.5 | 0.5 |
| Vancomycin | 100 | 1.0 | 1.0 |
| Daptomycin | 100 | 0.25 | 0.25 |
| TMP/SMX | 97.9 | ≤0.5 | ≤0.5 |
| **MRSA (573)** | | | |
| Delafloxacin | – | 0.06 | 0.5 |
| Levofloxacin | 30 | 4.0 | >4.0 |
| Ceftaroline | 95.3 | 1.0 | 1.0 |
| Clindamycin | 77.5 | ≤0.25 | >2.0 |
| Vancomycin | 100 | 1.0 | 1.0 |
| Daptomycin | 99.5 | 0.25 | 0.5 |
| TMP-SMX | 97.9 | ≤0.5 | ≤0.5 |
| ***E. faecalis*** (450) | | | |
| Delafloxacin | – | 0.06 | 1.0 |
| Levofloxacin | 70.7 | 1.0 | >4.0 |
| Ceftaroline | – | 1.0 | 1.0 |
| Vancomycin | 97.8 | 1.0 | 2.0 |
| Daptomycin | 100 | 1.0 | 2.0 |
| TMP/SMX | – | ≤0.5 | ≤0.5 |
| ***S. pyogenes*** (433) | | | |
| Delafloxacin | – | 0.008 | 0.015 |
| Levofloxacin | 99.8 | 0.5 | 1.0 |
| Ceftaroline | 100 | ≤0.015 | ≤0.015 |
| Vancomycin | 100 | 0.25 | 0.5 |
| Clindamycin | 91.5 | ≤0.25 | ≤0.25 |
| ***S. agalactiae*** (225) | | | |
| Delafloxacin | – | 0.008 | 0.015 |
| Levofloxacin | 97.8 | 0.5 | 1.0 |
| Ceftaroline | 100 | ≤0.015 | 0.03 |
| Vancomycin | 100 | 0.5 | 0.5 |
| Clindamycin | 70.7 | ≤0.25 | >2.0 |

Data obtained from ref. 10

Abbreviations: *MSSA* methicillin-susceptible *S. aureus*, *MRSA* methicillin-resistant *S. aureus*

**Table 5.3** In vitro activities of delafloxacin and comparators against gram-negative pathogens

| Organisms (no.)/antibiotic | % Susceptible | MIC-50 | MIC-90 (μg/mL) |
|---|---|---|---|
| ***Enterobacteriaceae*** (2250) | | | |
| Delafloxacin | – | 0.06 | 4.0 |
| Ciprofloxacin | 81.6 | ≤0.03 | >4.0 |
| Ceftriaxone | 80.3 | 0.12 | >8.0 |
| Ceftazidime | 86.3 | 0.25 | 16.0 |
| Pip./Tazo. | 89.3 | 2.0 | 32.0 |
| *E. coli* (500) | | | |
| Delafloxacin | – | 0.03 | 4 |
| Ciprofloxacin | 69.4 | ≤0.03 | >4 |
| Ceftriaxone | 84.0 | ≤0.06 | >8 |
| Ceftazidime | 89.2 | 0.12 | 8.0 |
| Pip./Tazo. | 94.2 | 2.0 | 8.0 |
| *K. pneumonia* (389) | | | |
| Delafloxacin | – | 0.06 | >4 |
| Ciprofloxacin | 77.4 | ≤0.03 | >4 |
| Ceftriaxone | 75.3 | ≤0.06 | >4 |
| Ceftazidime | 76.9 | 0.12 | >32 |
| Pip./Tazo. | 81.2 | 4.0 | >64 |
| *P. aeruginosa* (200) | | | |
| Delafloxacin | – | 0.25 | >4 |
| Ciprofloxacin | 75 | 0.25 | >4 |
| Ceftazidime | 78.5 | 2.0 | >32 |
| Pip./Tazo. | 81.2 | 8.0 | >64 |

Resistance of gram-negative bacteria to delafloxacin is similar to other FQs with cross-resistance. *E. coli* isolates with resistance to ciprofloxacin and delafloxacin usually demonstrate triple mutations in the QRDR [13]. In *Neisseria gonorrhoeae* isolates resistant to ciprofloxacin, reduced susceptibility to delafloxacin has been demonstrated with triple mutations in *gyrA* and *parC* [14]. Reduced susceptibility to delafloxacin can occur with certain single point mutation in *gyrA* and mutations in *mtrE* and *norM* efflux pumps.

## 5.5 Clinical Efficacy

Data on the clinical efficacy of delafloxacin were largely derived from studies on ABSSSIs and to a lesser extent on CAP. Two phase II trials in ABSSSI were performed with intravenous (IV) delafloxacin at 2 different doses, 300 and 450 mg, in the first study compared to IV tigecycline 100 mg daily for 1 week in 150 adults with complicated mainly MRSA infections [1]. The cure rates were not significantly different. The second study compared IV delafloxacin 300 mg, 600 mg linezolid, and vancomycin 15 mg/kg twice daily for a week in 250 adults with predominantly MRSA infections [1]. The clinical cure rates were similar between all groups in patients infected with MRSA.

Two phase III multicenter, randomized, double-blind trials of 1560 adults with ABSSSI were conducted, and the pooled data has been reported [15]. In one study (#302) IV delafloxacin 300 mg twice daily was compared to IV vancomycin 15 mg/kg twice daily plus 1–2 g of aztreonam twice daily for 5–14 days. In the second study (#303), delafloxacin 300 mg twice daily was given for 3 days with a switch to oral 450 mg twice daily compared to vancomycin plus aztreonam with the same dosing as study 302 for 4–14 days. The aztreonam could be discontinued if gram-negative pathogens were not isolated. These infections included cellulitis, abscesses, wound infections, and infected burns. *S. aureus* was the most common pathogen with 44% due to MRSA. The primary endpoint was response at 48–72 h with ≥20% reduction of lesion size, and the secondary endpoints was resolution of signs and symptoms at follow-up 14 days and 21–28 days. Overall, IV/oral delafloxacin was noninferior to vancomycin/aztreonam therapy with resolution of signs and symptoms at 14 days (84.7% and 84.1%) or 21–28 days (82% and 81.7%). The rates of eradication of MRSA were also similar (98.1% and 98%) and adverse events were comparable [15].

A global multicenter, randomized, double-blind, comparator-controlled, phase III trial with delafloxacin versus moxifloxacin (IV/oral) for treatment of CAP (DEFINE-CARBP) was recently reported [16]. Delafloxacin 300 mg twice daily was compared to moxifloxacin 400 mg daily for 5–10 days, and subjects could be switched to oral therapy after six IV doses. The study enrolled 859 subjects, 431 randomized (stratified by risk class) to delafloxacin and 428 to moxifloxacin. Surprisingly, 60.5% had at least one pathogen identified, *S. pneumoniae* (43.5%), *Haemophilus parainfluenzae* (14.6%), *Mycoplasma pneumoniae* (12.5%), *L. pneumophila* (11.9%), *H. influenzae* (11.9%), *S. aureus* (11.0%) with only 2 MRSA isolated, *Chlamydia pneumoniae* (7.9%), *K. pneumoniae* (6.3%), *E. coli* (5.2%), and *P. aeruginosa* (4.6%).

The early clinical response at 96 h and test of cure at 5–10 days post-therapy were similar between the two groups, 88.9% for delafloxacin and 89.0% for moxifloxacin. However, subjects with chronic obstructive pulmonary disease (COPD) or asthma were significantly better with delafloxacin than moxifloxacin (93.4% vs 76.8%) [16]. Treatment-related adverse events were comparable between delafloxacin (15.2%) and moxifloxacin (12.6%).

## 5.6 Adverse Events

Delafloxacin appears to be well tolerated and the most common side effects were mild-moderate gastrointestinal (GI) events (8% nausea or 8% diarrhea) [16]. Three patients in pooled phase 3 studies reported mild-moderate tendinitis [16]. In the DEFINE-CARBP trial, 2.1% of subjects discontinued delafloxacin due to treatment-related adverse events, but liver disturbance (primarily increased transaminases) were higher (5.1%) than with moxifloxacin (2.8%) [15]. FQs are considered high risk for development of *Clostridium difficile* colitis, but delafloxacin has greater activity against this pathogen than other FQs. However, two patients from phase II and III studies combined had *C. difficile* diarrhea [17]. Central nervous system (CNS)-related adverse events (headaches, dizziness, acute psychosis, or seizures) are the second most commonly reported side effects with FQ. This has been attributed to FQ blockade of GABA receptors, but in experiments the concentration of delafloxacin to inhibit these receptors is much higher than clinically achievable [16]. In pooled data from phase III studies of delafloxacin, headaches occurred in 3% and seizures was not reported except for one patient in a phase II study given 450 mg IV [17]. To date the use of delafloxacin has been limited, but the FDA warning on serious adverse events associated with the FQ class should be considered before treatment with this agent.

## 5.7 Role in Therapy

Delafloxacin would be most suitable in the treatment of complicated infections with mixed *S. aureus* (especially MRSA) and gram-negative pathogens (e.g., skin/soft tissue or wound infections), where one agent can be used IV in hospital or oral as outpatient therapy and avoid the use of combination agents. Comparative cost-benefit studies would be beneficial. Based on in vitro data, delafloxacin should be very effective in biofilm infection (e.g., chronic osteomyelitis, prosthetic infections, etc.), but clinical studies would be needed to confirm its efficacy over other agents; this could also apply to intracellular bacterial infections.

Because of the FDA black box warnings, older patients or patients with conditions that predispose to atherosclerotic disease and cardiac disease or CNS diseases should be avoided, and long-term therapy should be restricted to avoid tendinopathy and neuropathy. Restriction of the use of FQs in general has been implemented in antibiotic stewardship programs to slow the spread of global antimicrobial resistance pandemic.

## References

1. Mogle BT, Steele JM, Thomas SJ, Bohan KB, Kufel WD. Clinical review of delafloxacin: a novel anionic fluoroquinolone. J Antimicrob Chemother. 2018;73:1439–51.
2. Scott LJ. Delafloxacin: a review in acute bacterial skin and skin structure infections. Drugs. 2020;80:1247–58.
3. Hoover R, Hunt T, Benedict M, et al. Single and multiple ascending-dose studies of oral delafloxacin: effects of food, sex, and age. Clin Ther. 2016;38:39–52.
4. Hoover R, Hunt T, Benedict M, et al. Safety, tolerability, and pharmacokinetic properties of intravenous delafloxacin after single and multiple doses in healthy volunteers. Clin Ther. 2016;38:53–65.

5. Nilius AM, Shen LL, Hensey-Rudloff D, et al. *In vitro* antibacterial potency and spectrum of ABT-492, a new fluoroquinolone. Antimicrob Agents Chemother. 2003;47:3260–9.
6. Burak ES, Devito JA, Remy JM, et al. Delafloxacin chemical properties lead to increased potency against gram-positive pathogens, including quinolone-resistant pathogens. In: Proceedings of the 50th intersc conf antimicrob agents chemother; Boston, MA, USA, 20210. {Poster E-182}. Washington, DC: American Society for Microbiology.
7. Lemaire S, Tulkens PM, Bambbeke VF. Contrasting effects of acidic pH on the extracellular and intracelllular activities of the anti-gram-positive fluoroquinolones against *Staphylococcus aureus*. J Antimicrob Chemother. 2011;55:649–58.
8. Bauer J, Siala W, Tulkens PM, Van Bambeke F. A combined pharmacodynamics quantitative and qualitative model reveals the potent activity of daptomycin and delafloxacin against *Staphylococcus aureus* biofilms. Antimicrob Agents Chemother. 2013;57:2726–37.
9. Pfaller MA, Flamm RK, McCurdy SP, Pillar CM, Shortridge D, Jones RN. Delafloxacin *in vitro* broth microdilution and disk diffusion antimicrobial susceptibility breakpoint criteria and quality control ranges for an expanded-spectrum anionic fluoroquinolone. J Clin Microbiol. 2018;56:e00339–18.
10. Tulkens PM, Bamabeke F, Zinner SH. Profile of a novel fluoroquinolone—delafloxacin. Clin Infect Dis. 2019;68(S3):S213–22.
11. Brown Elliot BA, Wallace RJ. Comparison of *in vitro* susceptibility of delafloxacin with ciprofloxacin, moxifloxacin, and other comparator antimicrobials against isolates on nontuberculous mycobacteria. Antimicrob Agents Chemother. 2021;65(7):e0007921. https://doi.org/10.1128/AAC.00079-21.
12. Hermsen ED, Hovde LB, Konstantinides GN, et al. Mutant prevention concentrations of ABT-492, levofloxacin, moxifloxacin, and gatifloxacin against three common respiratory pathogens. Antimicrob Agents Chemother. 2005;49:1633–5.
13. Armer LS, Hoffrage JB, Keller EL, et al. In vitro bactericidal activities against gram-positive and gram-negative organisms. Antimicrob Agents Chemother. 2004;49:2771.
14. Soge OO, Salipante SJ, No D, et al. In vitro activity of delafloxacin against clinical *Neisseria gonorrhoeae* isolates and selection of gonococcal delafloxacin resistance. Antimicrob Agents Chemother. 2016;60:106–11.
15. Giordano PA, Pogue JM, Cammarata S. Analysis of pooled phase III efficacy data for delafloxacin in acute bacterial skin and skin structure infections. Clin Infect Dis. 2019;68(S3):S223–32.
16. Horcajada JP, Salata RA, Alvarez-Sala R, et al. A phase 3 study to compare delafloxacin with moxifloxacin for the treatment of adults with community-acquired bacterial pneumonia [DEFINE-CABP]. Open forum. Infect Dis. 2020;7:ofz514. https://doi.org/10.1093/ofid/ofz514.
17. Jorgensen SCJ, Mercuro NJ, Davis SL, Ryback MJ. Delafloxacin: place in therapy and review of microbiologic, clinical and pharmacologic properties. Infect Dis Therapy. 2018;7:197–217.

# Chapter 6
# New Oxazolidinone: Tedizolid

## 6.1 Introduction

Oxazolidinones are synthetic antimicrobials developed over 30 years ago with approval of linezolid for treatment of complicated skin and soft tissue infection and community-acquired pneumonia due to multiresistant gram-positive bacteria in 2000 in the United States (US). The oxazolidinones produce bacteriostatic effect by inhibiting protein synthesis by binding of the 23S ribosomal RNA of the 50S subunit [1]. The second-generation oxazolidinone, tedizolid phosphate (Sivextro), was approved in the US for treatment of acute bacterial skin and skin structure infections (ABSSSI) in 2014. The prodrug tedizolid phosphate is cleaved in the blood by serum phosphatase to form the active compound—tedizolid. The phosphate group improves water solubility and oral bioavailability and prevents the C-5 hydroxymethyl side chain from interactions with monoamine oxidase (MAO) [1]. The structure of tedizolid also differs from linezolid by the addition of a fourth para-orientated ring structure (D-ring in Fig. 6.1), which stabilizes interactions with target receptors and may be responsible for the increased activity above linezolid [1].

## 6.2 Antimicrobial Activity

Tedizolid has similar antimicrobial spectrum as linezolid: gram-positive bacteria—staphylococci, streptococci, enterococci, micrococci, *Bacillus* spp., *Corynebacterium* spp., *Listeria monocytogenes*, and some gram-positive anaerobes, *Mycobacteria tuberculosis (including multiresistant strains), non-tuberculous mycobacteria* (i.e., *M. avium* complex, *M. abscessus*), and *Nocardia* species [2–4]. Tedizolid is more potent than linezolid against gram-positive bacteria and the susceptibility of 6884 isolates from the US and European countries showed that the $MIC_{90}$ was fourfold



I. W. Fong, *New Antimicrobials: For the Present and the Future*, Emerging Infectious Diseases of the 21st Century,
https://doi.org/10.1007/978-3-031-26078-0_6

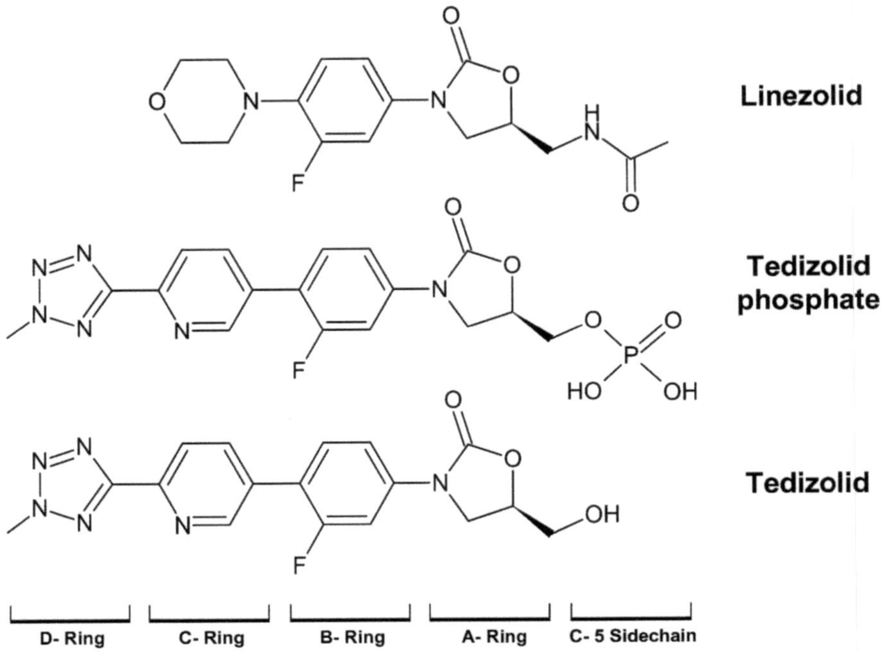


**Fig. 6.1** Diagram of structures of linezolid, tedizolid, and the phosphate prodrug. (Adopted from Ref. [1])

lower than those of linezolid; see Table 6.1 [1]. These include methicillin-resistant *Staphylococcus aureus* (MRSA) and vancomycin-resistant enterococci (VRE); 99.8% of isolates were susceptible. Strains with resistance or nonsusceptible to linezolid may be susceptible to tedizolid but have elevated MIC (0.5–8 μg/ml). The FDA-approved breakpoint for tedizolid susceptibility for regular gram-positive bacteria is ≤0.5 μg/ml. The oxazolidinones have bacteria-static activity and are not recommended for treatment of bacterial endocarditis.

The oxazolidinones have a role in the therapy of multidrug-resistant (MDR) *M. tuberculosis* and nontuberculous mycobacteria (*M. avium* complex and *M. abscessus*) which are often resistant to standard therapy. Linezolid has been used successfully (90% response) with bedaquiline and pretomanid for 6 months for highly drug-resistant tuberculosis, but toxic effects of linezolid were common with peripheral neuropathy in 81% and myelosuppression in 48% [5]. Hence, tedizolid with better safety profile than linezolid could be used instead for prolonged therapy, as it is more active in vitro against these strains of mycobacteria. In a study of 120 *M. tuberculosis,* 59 with drug resistance and 25 MDR, tedizolid showed 1–three-fold more potency than linezolid with all inhibited at concentration of ≤0.5 μg/ml [6]. These results were similar to a study in 2006 of 95 *M. tuberculosis* with 25 MDR isolates, all isolates were inhibited by ≤0.5 μg/ml of tedizolid, including those with MDR, and linezolid $MIC^{90}$ were 2 μg/ml (four-fold less active) [7].

**Table 6.1** Comparative in vitro activity of oxazolidinones against common gram-positive bacteria

| Organisms | Linezolid | | Tedizolid | |
|---|---|---|---|---|
| (no.) | $MIC_{50}$ | $MIC_{90}$ | $MIC_{50}$ | $MIC_{90}$ |
| *S. aureus* | μg/ml | | μg/ml | |
| MSSA (2729) | 2 | 2 | 0.25 | 0.5 |
| MRSA (1770) | 2 | 2 | 0.25 | 0.5 |
| CNS (537) | 1 | 1–2[a] | ≤0.12 | 0.25 |
| *Enterococcus* spp. | | | | |
| *E. faecalis* (221) | 1 | 2 | 0.25 | 0.5 |
| *E. faecium* (634) | 2 | 2 | 0.25 | 0.5 |
| VRE (164) | 2 | 2 | 0.25 | 0.5 |
| VSE (705) | 1 | 2 | 0.25 | 0.5 |
| *Streptococcus* spp. | | | | |
| β-hemolytic strep. (975) | 1 | 1 | 0.12 | 0.25 |
| *S. pneumoniae* | 1 | 2 | 0.25 | 0.25 |

Data obtained from Ref. [1]

Abbreviations: *CNS* coagulase negative staphylococci, *MSSA* methicillin-susceptible *S. aureus*, *MRSA* methicillin-resistant *S. aureus*, *VRE* vancomycin-resistant enterococci, *VSE* vancomycin-susceptible enterococci

[a]$MIC_{90}$ 1 μg/ml for *S. epidermidis* and 2 μg/ml for other coagulase-negative *Staphylococcus* spp.

Tedizolid also showed good intracellular killing activity of *M. tuberculosis* within infected macrophages comparable with that of rifampin [8].

Linezolid is also sometimes used in the treatment of MDR nontuberculous mycobacteria (NTM) infections, especially *Mycobacterium abscessus* complex and *Mycobacterium avium* complex. In an in vitro study of 170 isolates of rapidly growing NTM, the $MIC_{50}$ and $MIC_{90}$ of tedizolid were 1–8 fold lower than linezolid [9]. The tedizolid $MIC_{90}$ of 81 isolates of *M. abscessus* subsp. *abscessus* and 12 isolates *of M. abscessus* subsp. *massiliense* were 8 μg/ml and 4 μg/ml, respectively, versus linezolid $MIC_{90}$ of 32 μg/ml [9]. These results are similar to another report published in 2018 [10] (see Table 6.2). For linezolid isolates with MICs ≤8 were classified as susceptible [11] and 52.3% of the *M. abscessus* complex were considered sensitive [10]. If MICs ≤8 μg/ml of tedizolid were used as the tentative breakpoint, then 100% of the *M. abscessus* complex isolates would be susceptible to the newer oxazolidinone. In a recent study from Beijing, China, of 170 rapidly growing mycobacteria, the MIC values of tedizolid were the lowest among 4 oxazolidinone, including linezolid, for *M. abscessus* and *M. massiliense* [12]. The oxazolidinones are less active against *M. avium* complex, and among 100 isolates, most were resistant to linezolid and 50% or more would be considered susceptible to tedizolid if the same breakpoint were used (see Table 6.2).

*Nocardia* species are rare causes of severe infection in immunocompromised subjects and linezolid has been used for moderate-severe infections, alone or in combination as these bacteria are universally susceptible to this agent. In a recent case series of 20 patients treated with linezolid with median duration of 28 days, 45% developed thrombocytopenia and 40% developed anemia [13]. Thus, tedizolid

**Table 6.2** In vitro activity of oxazolidinones against *M. tuberculosis* and *M. abscessus*

| Organisms | Linezolid | | Tedizolid | |
|---|---|---|---|---|
| (no.) | $MIC_{50}$ | $MIC_{90}$ | $MIC_{50}$ | $MIC_{90}$ |
| *M. tuberculosis*[a] | | | | |
| Susceptible (97) | 1 | 2 | 0.25 | 0.5 |
| Monoresistant (68) | 1 | 2 | 0.25 | 0. 5 |
| MDR (50) | 1 | 2 | 0.25 | 0.5 |
| *M. abscessus* complex[b] (223) | | | | |
| *M. abscessus* (124) | 8 | ≥32 | 1–4 | 4–8 |
| *M. bolletii* (5) | 32 | >32 | 4 | 4 |
| *M. massiliense* (92) | 8 | ≥32 | 1 | 4 |
| *M. avium* complex (100) | 32 | 64 | 8 | >32 |

Abbreviations: *MDR* multiple drug resistant
[a]Data obtained from Refs. [6, 7]
[b]Data obtained from Refs. [9, 10]

with a better safety profile could replace linezolid if its in vitro activity were similar or better. In a study of 31 *Nocardia brasiliensis* isolates, the in vitro activity of tedizolid was similar to linezolid with $MIC_{90}$ of 1 μg/ml for both agents [7].

## 6.3 Mechanisms of Resistance

Resistance to the oxazolidinones, linezolid, appears to be rare, but with increasing use in the past two decades, there have been increasing reports of resistance to linezolid among enterococci and staphylococcal species. Nosocomial outbreaks of linezolid-resistant bacteria have been reported from several countries including Spain, Mexico, and Brazil [14], and risk factors include previous hospitalization and prior linezolid and prolonged use ≥30 days or recurrent use. In a single medical center in Mexico, 50 case patients with linezolid-resistant *E. faecalis* were identified over 3 years [14].

The main mechanism of oxazolidinone resistance is point mutation of the 23S rRNA binding site and the ribosomal proteins L3 and L4 [1]. Since staphylococci and enterococci possess 4–6 copies of the 23S rRNA genes, multiple mutations must be acquired to result in in vitro resistance. However, a single G2576T mutation can result in resistance [15]. The G2576T 23S rRNA mutation is the most common conferring high level resistance in MRSA, multiple-drug resistant *Staphylococcus epidermidis* (MDRSE), and VRE with linezolid treatment failure [16]. The L3, L4, and L22 proteins are close to the linezolid binding site in the ribosomal peptidyl transferase center, and mutations in these proteins may decrease linezolid binding and result in resistance in staphylococci and enterococci [16].

Acquisition of the *chloramphenicol-florfenicol resistance (cfr)* gene is of concern as it is a mobile genetic element that can be transferred horizontally by different bacteria. The *cfr* gene encodes for the RNA methyltransferase which adds a second methyl group at A2503 of the 23S rRNA, normally occupied by the C-5 side chain of linezolid and increases the MIC 2–4 fold [1]. This Cfr-mediated linezolid resistance does not appear to confer resistance to tedizolid in vitro, as the hydroxymethyl side chain is smaller and more flexible than the side chain of linezolid [1]. However, isolates with the *cfr* and the *ermB* methyltransferase gene demonstrate multidrug resistance to all protein synthesis inhibitor antibiotics [1, 16].

Resistance to the oxazolidinones can also be acquired by the transferable ribosomal protection genes, *optrA* and *poxtA*, which are part of the ATP-binding cassette (ABC) superfamily of proteins that are associated with antimicrobial resistance [17]. The gene *optrA* bestow resistance to phenicols and oxazolidinones and *poxtA* produce resistance to phenicols, oxazolidinones, and tetracyclines [16]. These genes can be found in staphylococci and enterococci, including MRSA and VRE.

## 6.4 Pharmacology and Pharmacokinetics

Tedizolid phosphate is converted to the active form by serum phosphatase immediately after intravenous (IV) infusion [1] and the oral formulation is converted to tedizolid by the intestinal apical alkaline phosphatase [18]. The maximum serum concentration ($C_{max}$) after 200 mg IV was 2.6 μg/ml at 1 h post-infusion, and after oral administration of the same dose, the $C_{max}$ was 2.0 μg/ml at 2 h with an elimination half-life of 12 h [1]. The bioavailability of oral tedizolid is about 92% and somewhat lower in Orientals (83–86%) with serum protein binding of 70–90% and larger volume of distribution than linezolid of 67–80 liters [18]. Metabolism of tedizolid occurs mainly in the liver and the inactive metabolite, tedizolid sulfate, is excreted in the feces (80%) and the urine (18%) [18]. No dosage adjustment is needed for renal or hepatic dysfunction, although severe hepatic dysfunction can result in 34% higher area under the 24 h concentration curve ($AUC_{0-24}$) [1].

Animal models of infection indicate that tedizolid antimicrobial activity is best correlated with the AUC:MIC ratios. A 200 mg daily dose of tedizolid is estimated to result in free $AUC_{0-24}$:MIC of about 3 μg/ml h/L, and in the presence of granulocytes, this could reduce MRSA bacterial burden by 3.5 $\log_{10}$ CFU at 48 h [19]. This was supported by phase II and phase III clinical trials of acute bacterial skin and skin structure infections (ABSSIs).

## 6.5 Clinical Activity

Initially tedizolid was assessed in a phase II, randomized, double-blind, dose-ranging study of complicated ABSSSI to receive 200 mg, 300 mg, or 400 mg daily for 5–7 days [20]. The clinical cure rates were similar between the treatment groups, 94.4–98.2%.

Subsequently, two phase III clinical trials, ESTABLISH -1 and ESTABLISH-2, were conducted to compare a 5-day course of tedizolid phosphate to 10-day course of linezolid in patients with ABSSSI. These randomized, multicenter, international, double-blind, noninferiority trials, stratified by clinical syndrome and geographic region, compared oral tedizolid 200 mg daily for 5 days versus linezolid 600 mg twice daily for 10 days. ESTABLISH-1 evaluated only oral therapy and ESTABLISH-2 patients were given two or more IV doses prior to having the option of switching to oral therapy [21, 22].

In ESTABLISH-1 trial, 332 patients received tedizolid and 336 received linezolid, and the primary response rates at 48–72 h met noninferiority criteria, 79.5% and 79.4%. The clinical response rates 7–14 days post-treatment were also very similar in patients with MRSA infection, 85.2% and 85.6% [21]. In ESTABLISH-2 trial, 332 patients received tedizolid and 334 linezolid, with similar early clinical response of 85% and 83%, respectively [22]. Pooling the data of both studies also confirmed noninferiority of the early response rate and the clinical response post-therapy evaluation 7–14 days after end of therapy, with 86.7% response for tedizolid and 86.8% for linezolid. These results are similar to a phase 3 randomized, multicenter, open-label study in Japan with smaller sample size ($n = 125$), but drug-related adverse events were lower with tedizolid (30.1%) than linezolid (39.0%), mainly gastrointestinal (21.7% vs 26.8%) and myelosuppression-related (2.4% vs 22.0%) [23].

A systematic review and meta-analysis of 15 trials with 3615 patients with ABSSSI caused by MRSA were evaluated for relative effectiveness of tedizolid and monotherapy comparators (ceftaroline, daptomycin, linezolid, teicoplanin, tigecycline, and vancomycin) [24]. Tedizolid was superior to vancomycin but similar to other comparators, and discontinuation due to adverse events was similar among the agents. Thus, tedizolid is an alternative option for the treatment of serious skin and soft tissue infections due to MRSA. This review, however, did not provide a cost comparison between the various therapies.

A recent phase 3, randomized, double-blind trial compared tedizolid and linezolid for gram-positive nosocomial pneumonia, including ventilator-associated pneumonia [25]. Tedizolid 200 mg daily for 7 days was given to 366 patients and linezolid 600 mg twice daily to 360 patients. The primary endpoint of all-cause mortality at 28 days was similar, 28.1% and 26.4%, respectively, and achieved noninferiority. The assessment of clinical response at test of cure was lower with tedizolid, 56.3% vs linezolid 63.9%, and noninferiority for tedizolid was not achieved. Drug-related adverse events occurred in 8.1% and 11.9% of patients receiving tedizolid and linezolid, respectively [23].

### 6.5.1 *Safety*

In the phase III trials, side effects were mild for both drugs with only 0.5% of patients on tedizolid and 0.9% on linezolid discontinuing therapy because of adverse events. The most common side effects were gastrointestinal in nature (mainly nausea) and occurred more often with linezolid, 23% versus 16%. In these short-term studies, hematological adverse events were also mild, such as mild thrombocytopenia (<150,000 platelets/mm$^3$) at 11–13 days (4.9% for tedizolid vs 10.8% for linezolid, $p = 0.0003$), mild anemia (28.9% with tedizolid and 31.1% for linezolid), and absolute neutrophil count just below lower limit of normal (1.9% for tedizolid vs 3.3% for linezolid) [1]. Duration of therapy in these trials were not long enough to detect neurological adverse events reported with linezolid, such as peripheral and optic neuropathies. Post-marketing monitoring for these adverse effects with tedizolid will be necessary to obtain an accurate estimate of these uncommon side effects seen mainly with longer-term therapy. This is true also for the serotonin syndrome reported with linezolid from interactions with monoamine oxidase (MAO) inhibitors (which were excluded in these trials), although animal studies suggest that tedizolid has low potential for causing MAO-related toxicity [1].

## 6.6 Place in Therapy

Theoretically tedizolid could replace linezolid for short-term treatment of complicated skin and skin structure infections due to MRSA and long-term treatment of MRSA prosthetic infections and MDR tuberculosis or nontuberculous mycobacterial infections. Its advantages over linezolid include once daily dosing, shorter duration for ABSSSIs, potentially less hematological and neurological adverse effects with long-term therapy, and possibly less drug-drug interactions with MAO inhibitors. The safety advantage of tedizolid versus linezolid may be greatest for long-term therapy where serious side effects are more commonly reported. However, data of the safety of tedizolid with long-term treatment is still limited and reports to date include small number of patients.

Tedizolid was used for 18 months in a single patient as suppressive therapy for recurrent MRSA for a peripheral vascular graft infection without hematological or neurological toxicity [26]. It has also been used in 24 patients with NTM infections (20 with pulmonary and 4 with disseminated disease) for median duration of 101 days (range, 15–369 days) [27]. Adverse events occurred in 14 (58%) with peripheral neuropathy in 5 (21%), muscle rigidity in 3 (13%) with metoclopramide suggestive of serotonin toxicity, thrombocytopenia in 1 and anemia in 1, liver enzyme abnormalities in 2 (8%), and gastrointestinal disturbance (nausea, vomiting, or diarrhea) in 6 (26%). In a case report, a patient with multiple myeloma and nocardia infection of the brain was treated with tedizolid 200 mg daily and sulfamethoxazole/trimethoprim (SMX/TMP) for 6 months with cure and no significant adverse events [28]. Tedizolid has also been used for 20 months with excellent tolerance in an adolescent with pulmonary tuberculosis after liver transplant, which was required due to hepatotoxicity from standard anti-tuberculous agents [29].

There appears to be a place for tedizolid in therapy of nocardiosis due to limited number of oral agents, need for long-term treatment, drug intolerance, resistant organisms, and toxicity. However, brain involvement is common in immunosuppressed patients, and the penetration of tedizolid in the central nervous system (CNS) has been questioned due to the high protein binding. So far 5 patients with *Nocardia* infection have been treated with tedizolid and SMX/TMP with success after failure of other combinations with SMX/TMP or toxicity, 2 with brain involvement [28, 30].

Tedizolid may have a role in the management of osteoarticular infections, especially with MRSA alone or in mixed infection with *Enterococcus* species, where linezolid may be considered. In a multicenter, retrospective study, 51 patients with osteoarticular infections were treated with tedizolid 200 mg daily for a median of 29 days [31]. Reasons for choosing tedizolid were drug-drug interactions with other agents (63%) and cytopenia (55%). Fifty-nine percent of the patients were orthopedic device-related with 17 (33.3%) with prosthetic joint infection and 24% of the patients received concomitant rifampin. Implant was retained in 33% of the devise-related infections, and the overall cure rate was 83% with a median follow-up of 630 days. Treatment with tedizolid was well tolerated and only 3 patients (6%) had gastrointestinal side effects (nausea and occasional vomiting). Three patients were switched from linezolid because of myelotoxicity and completed therapy with tedizolid without additional worsening [31].

Studies on the safety and efficacy of tedizolid in combination with new anti-tuberculous agents are warranted for MDR tuberculosis, and it probably could replace linezolid in the combination with bedaquiline and pretomanid for 6 months.

## References


1. Rybak JM, Roberts K. Tedizolid phosphate: a next-generation oxazolidinone. Infect Dis Ther. 2015;4:1–4.
2. Tsuji BT, Kaatz GW, Rybak MJ. Linezolid and other oxazolidinones. Infect Dis Antimicrob Agents; www.antimicrobe.org/new/d13.asp
3. Zhanel GG, Shroeder C, Vercaigne L, Gin AS, Embil J, Hoban DJ. A critical review of oxazolidinones: an alternative or replacement for glycopeptides and streptogramins? Can J Infect Dis. 2001;12:379–90.
4. Davidson N, Grigg MJ, Mcguiness SL, Baird RJ, Anstey NM. Safety and outcomes of linezolid use for nocardiosis. Open Forum Infect Dis. 2020;7:ofaa090.
5. Conradie F, Diacon AH, Ngubane N, et al. Treatment of highly drug-resistant pulmonary tuberculosis. N Engl J Med. 2020;382:893–902.
6. Ruiz P, Causse M, Vaquero M, Casai M. In vitro activity of tedizolid against *mycobacterium tuberculosis*. Antimicrob Agents Chemother. 2019;63:e01939–18.
7. Vera-Cabrera L, Gonzalez E, Rendon A, et al. In vitro activities of DA-7157 and Da-7218 against *Mycobacterium tuberculosis* and *Nocardia brasiliensis*. Antimicrob Agents Chemother. 2006;50:3170–2.
8. Molina-Torres CA, Barba-Marines A, Valles-Guerra O, et al. Intracellular activity of tedizolid phosphate and ACH-702 versus M*ycobacterium tuberculosis* infected macrophages. Ann Clin Microbiol Antimicrob. 2014;13:13. https://doi.org/10.1186/1476-0711-13-13.
9. Brown-Elliott B, Wallace RJ Jr. In vitro susceptibility of tedizolid against nontuberculous mycobacteria. J Clin Microbiol. 2017;56:JCM.00274-17.

10. Tang YW, Cheng B, Yeoh SF, Lin RTP, Teo JWP. Tedizolid activity against clinical *Mycobacteria abscessus* isolates—an *in vitro* characterization study. Front Microbiol. 2018;9:2095. https://doi.org/10.3389/fmib.2018.02095.
11. Woods GL, Brown-Elliott BA, Conville PS, Desmond EP, Hall GS, Lin G. Susceptibility testing of mycobacteria, nocardia, and other aerobic actinomycetes; approved standard-second edition. Clin Lab Stand Inst. 2011;26:1–61.
12. Wen S, Gao X, Zhao W, et al. *In vitro* efficacy comparison of linezolid, tedizolid, sutezolid, and delpazolid against rapid growing Mycobacteria isolated in Beijing, China. bioRxiv. https://doi.org/10.1101/2020.06.25.172742v1.full.
13. Davidson N, Grigg MJ, Mcguinness SL, Baird RJ, Anstey NM. Safety and outcome of linezolid use for nocardiosis. Open forum Infect Dis. 2020;7:ofaa090. https://doi.org/10.1093/ofid/ofaa090.
14. Rodriguez-Noriega E, Hernandez-Morfin N, Garza-Gonzalez E, et al. Risk factors and outcome associated with the acquisition of linezolid-resistant *Enterococcus faecalis*. J Glob Antimicrob Reisist. 2020;21:405–9.
15. Besier S, Ludwig A, Zander J, Brade V, Wichelhaus TA. Linezolid resistance in *Staphylococcus aureus*: gene dosage effect, stability, fitness costs, and cross resistance. Antimicrob Agents Chemother. 2008;52:1570–2.
16. Turner AM, Lee JYH, Gorrie CL, Howden BP, Carter GP. Genomic insights into last-line antimicrobial resistance in multidrug resistant *Staphylococcus aureus* and vancomycin-resistant *enterococcus*. Front Microbiol. 2021;12 https://doi.org/10.3389/micb.2021.6537656.
17. Jones PM, O'Mara ML, George AM. ABC transporters: a riddle wrapped in a mystery inside an enigma. Trends Biochem Sci. 2009;34:520–31.
18. Bassetti M, Castaldo N, Peghin M, Giacobbe DR. Tedizolid phosphate for the treatment of acute bacterial skin and skin- structure infections: an evidence-based review of its place in therapy. Core Evidence. 2019;14:31–40.
19. Lodise TP, Drusano GL. Use of pharmacokinetic/pharmacodynamics systems analyses to inform dose selection of tedizolid phosphate. Clin Infect Dis. 2014;58(Suppl 1):S28–34.
20. Prokocimer P, Bien P, Surber J, et al. Phase 2, randomized, double-blind, dose-ranging study evaluating the safety, tolerability, population pharmacokinetics and efficacy of oral tedizolid phosphate in patients with complicated skin and skin structure infections. Antimicrob Agents Chemother. 2011;55:583–92.
21. Prokocimer P, De Anda C, Fang E, Mehra P, Das A. Tedizolid phosphate vs linezolid for treatment of acute bacterial skin and skin structure infections: the ESTABLISH-1 randomized trial. JAMA. 2013;309:559–69.
22. Moran GJ, Fang E, Corey GR, Das AF, De Anda C, Prokocimer P. Tedizolid for 6 days versus linezolid for 10 days for acute bacterial skin and skin structure infections [ESTABLISH-2]: a randomized, double-blind, phase 3, non-inferiority trial. Lancet Infect Dis. 2014;309(14):696–705.
23. Mikamo H, Takesue Y, Iwamoto Y, Tanigawa T, Kato M, Tanimura Y, Kohno S. Efficacy, safety and pharmacokinetics of tedizolid versus linezolid in patients with skin and soft tissue infections in Japan—results of a randomized, multicenter phase 3 study. J Infect Chemother. 2018;24:434–42.
24. McCool R, Gould IM, Eales J, et al. Systemic review and network meta-analysis of tedizolid for treatment of acute bacterial skin and skin structure infections caused by MRSA. BMC Infect Dis. 2017;17:39.
25. Wunderink RG, Roquilly A, Croce M, et al. A phase 3, randomized, double-blind study comparing tedizolid phosphate and linezolid for treatment of ventilated gram-positive hospital-acquired or ventilator-associated bacterial pneumonia. Clin Infect Dis. 2021:ciab032. https://doi.org/10.1093/cid/ciab032.
26. Nigo M, Luce AM, Arias CA. Long-term use of tedizolid as suppressive therapy for recurrent methicillin-resistant *Staphylococcus aureus* graft infection. Clin Infect Dis. 2018;66:1975–6.

27. Kim T, Wills A, Markus A, Prevots DR, Olivier KN. Safety and tolerability of long term use of tedizolid for treatment of nontuberculous mycobacterial infections. Open Forum Infect Dis. 2016;3(Suppl. 1):577. https://doi.org/10.1093/ofid/ofw172.440.
28. Matin A, Sharma S, Mathur P, Apewokin SK. Myelosuppression-sparing treatment of central nervous system nocardiosis in a multiple myeloma patient utilizing a tedizolid-base regimen: a case report. Int J Antimicrob Agents. 2017;49:488–92.
29. Efficacy and safety of long-term use of tedizolid after liver transplantation in an adolescent with pulmonary tuberculosis. J Antimicorb Chemother. 2019;74:2817–9.
30. Odrobina R, Tritle B, Gomez CA. Successful use of tedizolid for the treatment of nocardiosis in transplant recipients. Am Transplant Congress. 2020; May 30: abstract no.: D-160.
31. Benavent E, Morata L, Escrihuel-Vidal F, et al. Long-term use of tedizolid in osteoarticular infections: benefits among oxazolidinone drugs. Antibiotics. 2021;10:53. https://doi.org/10.3390/antibiotics10010053.

# Chapter 7
# New Tetracyclines: Eravacycline and Omadacycline

## 7.1 Introduction

Tetracycline development began in the 1940s with the discovery of the natural antibiotic, chlortetracycline, produced by *Streptomyces* species. Shortly after, other naturally occurring tetracyclines were found, oxytetracycline, tetracycline, and demeclocycline. Tetracycline compounds consist of four cyclic hydrocarbon rings, which coined the class name "tetracycline," and they differ from each other by the presence of chloride, methyl, and hydroxyl groups [Wikipedia]. These modifications do not change their antimicrobial activity but influence their pharmacological properties (half-life, protein binding, etc.). Semisynthetic tetracycline was first produced in the 1950s, and by the 1960s to 1970s, the second-generation semisynthetic agents, doxycycline and minocycline, were produced [1].

Tetracyclines have a broad spectrum of activity, more than most classes of antibiotics, with activity against gram-positive and gram-negative bacteria, chlamydiae, mycoplasmas, rickettsiae, and protozoan parasites [2]. Widespread use of tetracyclines in humans and animals over the years has led to increased global resistance. The tetracyclines are bacteriostatic agents, growth inhibitors that are only effective against multiplying microbes. They diffuse passively through the porin channels in the bacterial membrane and bind reversibly to the bacterial 30S ribosomal unit to inhibit protein synthesis [2]. They can bind to the bacterial 50S ribosomal subunit and may alter cytoplasmic membrane to cause leakage of intracellular contents.

Resistance to tetracyclines may occur by efflux, ribosomal protection, reduced permeability, ribosomal mutation, and enzymatic inactivation (the rarest type). In the early 2000s, minocycline was structurally altered to produce tigecycline (a glycylcycline), to overcome resistance with expanded activity. However, high gastrointestinal (GI) toxicity and the Food and Drug Administration (FDA) warning of increased mortality compared to other comparators led to curtailed use. In 2018, two new tetracyclines, eravacycline and omadacycline, were approved for clinical



I. W. Fong, *New Antimicrobials: For the Present and the Future*, Emerging Infectious Diseases of the 21st Century,
https://doi.org/10.1007/978-3-031-26078-0_7

use by the FDA. These new compounds were specifically designed to overcome tetracycline resistance with improved safety profile.

## 7.2 Eravacycline

Eravacycline (ERV) is a fully synthetic novel fluorocycline with the tetracycline core of the four cyclic rings (A, B, C, and D) with integration of the fluorine at the C-7 and a pyrrolidinoacetamido group at the C9 on the D ring [3]. The modifications produce enhanced activity against tetracycline-resistant bacteria, and the pyrrolidinoacetamido group results in increased ribosomal binding and steric hindrance to avoid ribosome protection-related tetracycline resistance [4].

## 7.3 Pharmacokinetics and Pharmacodynamics of Eravacycline

Pharmacokinetic (PK) studies of ERV has shown a half-life of about 20 h that increases with repeated doses, an average protein binding of 71.4–82.5%, volume of distribution of about 320.0 L, with a maximum plasma concentration ($C_{max}$) of 1.29 μg/ml at 30 min, and area under plasma concentration-time curve from zero to 12 h ($AUC\text{-}_{0-12}$) of 4.56 μg.h/ml with the approved intravenous (IV) dose of 1 mg/kg every 12 h [5]. The oral bioavailability of ERV is low and average about 28%, and the oral formulation is pending approval. ERV is metabolized in the liver mainly by CYP3A4 and FMO-mediated oxidation, and it is primarily excreted in the feces with minor renal elimination [5]. Dose adjustment is not needed in renal failure or mild-moderate hepatic failure but is required in severe hepatic failure (Child-Pugh C).

The pharmacodynamics parameters best associated with ERV efficacy is the AUC over the minimum inhibitory concentration (MIC) [5]. Multiple dosing studies over 10 days show some accumulation (45%) with steady state at 5 days. ERV is widely distributed in the body as reflected by the volume of distribution being greater than the normal extracellular fluid volume [3]. Tigecycline and ERV appear to have the widest tissue distribution among the members of the tetracycline class as reflected by the volume of distribution: tigecycline 6–9 L/kg, ERV 4 L/kg, omadacycline 2.6 L/kg, tetracycline 1.3 L/kg, and doxycycline 0.7 L/kg [5].

## 7.4 Microbial Activity of Eravacycline

ERV spectrum of antimicrobial activity is similar to other members of the tetracycline class, but it was designed to overcome the two main resistance mechanisms common to the class: ribosomal protection, commonly found in gram-positive bacteria, and active drug efflux, found commonly in both gram-positive and gram-negative bacteria [5]. ERV has potent broad-spectrum activity against aerobic and anaerobic gram-positive and gram-negative bacteria, except for *Pseudomonas aeruginosa* and *Burkholderia cenocepacia* [6].

Against gram-positive bacteria, ERV has excellent in vitro activity against methicillin-susceptible and methicillin-resistant *Staphylococcus aureus* (MRSA), coagulase-negative staphylococci (CNS), vancomycin-susceptible and vancomycin-resistant *Enterococcus faecalis* and *Enterococcus faecium* (VRE), penicillin-susceptible and penicillin-resistant *Streptococcus pneumoniae,* and other streptococci with $MIC_{90}$ ranging from 0.016 to 0.5 μg/ml [6]. The FDA recommended susceptibility breakpoint for *Enterococcus* spp., *S. aureus*, and *Streptococcus anginosus* group is ≤0.06 μg/ml [3].

ERV in vitro activity has been assessed on >13,000 gram-negative bacilli collected from 2013 to 2017 worldwide. The FDA recommended susceptibility breakpoint for *Enterobacteriaceae was* ≤ 0.5 μg/ml. At this breakpoint, 92.6% of the *Enterobacteriaceae* were susceptible: 98.8% of *Escherichia coli* (including extended-spectrum β-lactamase [ESBL] producers), 90.6% of *Klebsiella* spp., 89.6% of *Enterobacter* spp., 94.6% of *Citrobacter* spp., 70.5% of *Acinetobacter baumannii*, 41.7% of *Stenotrophomonas maltophilia*, 10.6% of *Proteus mirabilis*, 10.3% of *Serratia marcescens*, and 1.0% of *Pseudomonas aeruginosa* [4]. Among multidrug-resistant isolates, 80.5% of the *Enterobacteriaceae* were susceptible to ERV. The $MIC_{90}$ values for *Salmonella* spp., *Shigella* spp., *Haemophilus influenzae*, *Moraxella catarrhalis*, and *Acinetobacter lwoffii* were found to be ≤0.5 μg/ml [5]. For gram-positive cocci and gram-negative bacilli, ERV MICs are usually twofold less than tigecycline.

Among anaerobes, ERV demonstrated four- to eightfold greater activity than other tetracyclines and the FDA-approved susceptibility breakpoint for anaerobes was ≤0.5 μg/ml [3]. It is highly active against *Actinomyces* spp. (limited no. tested), *Anaerococcus* spp., *Clostridium difficile, Fusobacterium* spp., *Bacteroides vulgatus*, and *Bifidobacterium* spp. with $MIC_{90}$ usually ≤0.5 μg/ml, and somewhat less active against *Clostridium perfringens*, *Bacteroides fragilis*, and *Prevotella* spp. with $MIC_{90}$ 1–2 μg/ml [3, 6]. ERV also has activity against atypical organisms including *Chlamydophila pneumoniae*, *Mycoplasma pneumoniae*, and *Legionella* spp. with *Legionella pneumophila* $MIC_{90}$ of 1–2 μg/ml [5]. It also has good activity against *Mycobacterium abscessus* with $MIC_{50}/MIC_{90}$ of 0.5/1.0 μg/ml [7]. Tables 7.1 and 7.2 summarize the comparative in vitro activity of the new tetracycline agents and comparators.

## 7.5 Resistance Mechanisms Affecting Eravacycline

Of the four resistant mechanisms for tetracycline bacterial resistance, efflux and ribosomal protection proteins are the most prominent. Resistance to ERV and omadacycline has been rarely reported at present. ERV evades TEtA efflux pumps which

**Table 7.1** Comparative in vitro activity of the new tetracyclines against gram-positive bacteria

| Organisms | Tetracycline $MIC_{50}/MIC_{90}$ μg/ml | Tigecycline $MIC_{50}/MIC_{90}$ μg/ml | Eravacycline $MIC_{50}/MIC_{90}$ μg/ml | Omadacycline $MIC_{50}/MIC_{90}$ μg/ml |
|---|---|---|---|---|
| *S. aureus* | | | | |
| MSSA | ≤0.5/≤0.5 | 0.06/0.12 | 0.12/0.25 | 0.12/0.25 |
| MRSA | ≤0.5/4 | 0.06/0.12 | 0.12/0.25 | 0.12/0.25 |
| *E. faecalis* | >16/>16 | 0.06/0.12 | 0.06/0.25 | 0.12/0.25 |
| *E. faecium* | 16/>16 | 0.03/0.06 | 0.06/0.12 | 0.06/0.12 |
| *S. pneumoniae* | ≤0.25/>8 | 0.03/0.06 | 0.015/0.03 | 0.06/0.12 |
| *β-hem. Strep.* | 0.5/>8 | 0.06/0.06 | 0.15/0.03 | 0.06/0.12 |
| *S. anginosus* gp. | 0.5/>8 | 0.03/0.03 | — | 0.06/0.12 |
| *C. difficile* | — | 0.25/0.25 | 0.06/0.12 | 0.25/0.5 |
| *C. perfringens* | — | 8/>16 | 1/1.2 | 4/16 |
| *Peptostrep.* spp. | — | 0.12/2 | — | 0.12/1 |

Data obtained from Refs. [3, 14]
Abbreviations: *MSSA* methicillin-susceptible *S. aureus*, *MRSA* methicillin-resistant *S. aureus*, *β-hem. Strep.* β-hemolytic streptococci, *Peptostrep. Peptostreptococci*, *MIC* minimum inhibitory concentration

are responsible for a large proportion of tetracycline resistance in *Enterobacteriaceae*. However, *Klebsiella pneumoniae* has developed ERV resistance related to overexpression of the OqxAB and MacAB efflux pumps [8] and *E. coli* overexpressing the tetracycline (TET) degrading enzymes [3]. The Tet (x3) and Tet(x4) plasmid-encoded genes isolated from *Enterobacteriaceae* and *Acinetobacter* strains from human and animal sources encoding resistance to ERV and omadacycline are of concern [9]. Additionally, mutations of the ribosomal target from isolated mutants with increased ERV MICs contain mutations in rpsJ [9].

## 7.6 Clinical Efficacy and Safety of Eravacycline

Clinical efficacy of ERV was assessed in one phase 2 study and two phase 3 studies of intra-abdominal infections, predominantly due to complicated appendicitis. In the phase 2 randomized, controlled trial (RCT), two doses of ERV were assessed (IV ERV at 1.5 mg/kg every 24 h and 1 mg/kg every 12 h) compared to ertapenem 1 g every 24 h for 4–14 days [3]. Clinical success was achieved in 92.3% of ERV once daily, 100% of the ERV twice daily, and 92.3% in the ertapenem group. Adverse events were mainly nausea and vomiting and occurred in 35.8% (ERV once daily), 28.6% (ERV twice daily), and 26.7% in the ertapenem group [3].

**Table 7.2** Comparative in vitro activity of the new tetracyclines against gram-negative and atypical bacteria

| Organism | Tetracycline $MIC_{50}/MIC_{90}$ µg/ml | Levofloxacin $MIC_{50}/MIC_{90}$ µg/ml | Tigecycline $MIC_{50}/MIC_{90}$ µg/ml | Eravacycline $MIC_{50}/MIC_{90}$ µg/ml | Omadacycline $IC_{50}/MIC_{90}$ µg/ml |
|---|---|---|---|---|---|
| *E. coli* | 2/>16 | ≤0.06/4 | 0.12/0.25 | 0.12/0.5 | 0.5/2 |
| *K. pneumoniae* | 2/>16 | ≤0.06/32 | 0.25/1 | 0.25/2 | 2/8 |
| *E cloacae* spp. | 2/16 | — | 0.25/0.5 | 0.5/2 | 2/4 |
| *Citrobacter* spp. | 1/4 | — | 0.25/0.5 | 0.5/2 | 1/4 |
| *P. mirabilis* | >16/>16 | — | 2/4 | 2/4 | 16/32 |
| *H. influenzae* | 0.5/1 | ≤0.06/≤0.06 | 0.12/0.25 | 0.12/0.25 | 1/1 |
| *M. catarrhalis* | 0.25/0.5 | — | 0.06/0.06 | 0.03/0.06 | 0.25/0.25 |
| *B. fragilis* | — | — | 0.5/2 | 0.5/2 | 0.5/4 |
| *B. ovatus* | — | — | 0.5/8 | 0.5/8 | 0.5/4 |
| *B. vulgatus* | — | — | 0.25/1 | 0.12/0.5 | 0.25/1 |
| **Atypical bacteria**[a] | | | | | |
| *M. pneumoniae* | 025/0.5 | — | — | — | 0.12/0.25 |
| *L. pneumophila* | 1/1 | — | — | — | 0.25/0.25 |
| *C. pneumoniae* | 0.12/0.12 | — | — | — | 0.06/0.25 |
| *M. abscessus* | >64/>64 | — | 2/2 | 1/1 | 2/2 |
| *M. chelonae* | 32/64 | — | 0.06/0.25 | — | 0.12/0.25 |
| *M. fortuitum* | 8/64 | — | 0.25/0.5 | — | 0.12/0.5 |

Data obtained from Refs. [3, 14, 19]
[a]Tetracycline susceptibility was for doxycycline

The phase 3 IGNITE1 study was a noninferiority, double blind, multicenter RCT from 66 sites in the US, Argentina, and South Africa with complicated intra-abdominal infections (cIAI) in 541 patients randomized to ERV 1 mg/kg every 12 h or ertapenem 1 g every 24 h for 4–14 days [10]. Noninferiority criteria were met, and the clinical cure rates were similar: 87.0% for ERV and 88.8% for ertapenem. The distribution of ESBL-producing *Enterobacteriaceae* was similar in the two groups (12.5% for ERV and 10.5% for ertapenem), but low numbers of carbapenem-resistant isolates and *P. aeruginosa* were recovered.

A phase 3 study IGNITE4 compared IV ERV 1 mg/kg every 12 h to 1 g meropenem daily in a noninferiority, double-blind RCT in 500 patients with cIAI from 65 sites in the US and Europe [11]. The main underlying disease in both groups was complicated appendicitis. The response rates were similar with clinical cure of 92.4% for ERV and 91.6% for meropenem. Pooling of the data from the 3 RCTs showed no difference in the clinical response, but ERV was associated with greater odds of adverse events (nausea), but no difference in vomiting or serious adverse events or discontinuation due to side effects [12].

ERV was evaluated for complicated urinary tract infections (cUTIs) in the IGNITE2 and IGNITE3 phase 3 RCTs. In the IGNITE2, IV ERV 1.5 mg/kg every 24 h is compared to IV levofloxacin 750 mg every 24 h, with a minimum of the first three doses and then oral transition on adequate clinical improvement for a total of 7 days [3]. Two doses of oral ERV were evaluated, either 200 mg twice a day or 250 mg twice a day, but the oral dose of levofloxacin was 750 mg daily. Based on similar efficacy and lower side effects, the main portion of the study was continued with only ERV 200 mg twice a day. Although 908 patients were randomized, only 600 were included in the microbiological assessment. The UTI conditions included acute pyelonephritis (48.6%), obstructive uropathy (24.1%), and surgical and abnormal anatomy. Baseline pathogens were balanced between the two groups, but for the primary endpoint, ERV did not meet noninferiority compared to levofloxacin, 60.4% vs 66.9%, for composite of clinical and microbiological cure. Adverse events were higher in the ERV arm, 37.1% vs 22.7%.

IGNITE3 trial compared IV ERV 1.5 m/kg every 24 h to IV ertapenem 1 g every 24 h for 5 days with option of switching to oral levofloxacin 750 mg every 24 h for total of 7–10 days for cUTI. The primary outcome was the composite of clinical and microbiological cure. Again ERV did not meet noninferiority with microbiological response of 84.8% vs 94.8% for ertapenem [3].

In the studies to date, ERV adverse events have been mild with mainly infusion site reactions (9.3%), nausea (8.1%), vomiting 4.1%, and diarrhea (2.2%) and mild liver disturbance of <12–20%, similar to the comparators [3]. ERV is expected to have similar side effects to other tetracyclines which were not detected in these trials, and *C. difficile* colitis associated with ERV was not reported in any of the studies [3]. Drug-drug interaction can occur with CYP3A4 inducers (e.g., rifampin) lowering blood levels of ERV and CYP inhibitors (e.g., itraconazole) increasing blood concentration and AUC [3].

## 7.7 Omadacycline

Omadacycline (OMAD) was approved by the FDA in 2018 for the treatment of community-acquired bacterial pneumonia (CAP) and acute bacterial skin and skin structure infections (ABSSIs), marketed as Nuzyra. It is a semisynthetic aminomethylcycline derivative of minocycline with modifications at C-7 and C-9 of the tetracycline D-ring (similar to ERV), but with an aminomethyl group present at the C-9 position [13]. The C-7 modifications evade the tetracycline-specific efflux resistance mechanism and the C-9 modification overwhelms the ribosomal protection resistance mechanism [14].

### 7.7.1 Pharmacokinetics and Pharmacodynamics of Omadacycline

OMAD is available in IV and oral formulations with bioavailability of 34.5% leading to oral dose of 300 mg versus 100 mg IV dose [13]. The half-life is 16–17 h, protein binding of 20%, and volume of distribution 2.6 L/kg, lower than that of ERV. OMA undergoes minimal metabolism in the liver and is not affected by the cytochrome P450 system, but it is predominantly excreted by the feces (81.1%) and less by renal elimination (14.4%) [13]. Dose adjustment is not necessary for renal impairment or failure. The oral preparation should be taken on an empty stomach as dairy products and multivalent cations can impair absorption.

The pharmacodynamic properties of OMAD have been studied by in vitro and in vivo method. Although all tetracyclines mechanisms of action are similar and are bacteriostatic, OMAD has bactericidal or bacteriostatic activity that is organism dependent [14]. It has bactericidal activity (≥3 log reduction of inoculum) against *S. pneumoniae*, *H. influenzae*, and *M. catarrhalis*, but bacteriostatic activity against *S. aureus*, *enterococci*, and *E. coli* [14]. Like other members of the tetracycline class, the AUC/MIC ratio is the best pharmacodynamic measurement that predicts the efficacy of OMAD [15].

For gram-positive bacteria (*S. aureus, S. pneumoniae*, and enterococc*i*) and *E. coli*, OMAD has the post-antibiotic effect of 1.4–3.3 h and demonstrated intracellular killing of *S. aureus* (including MRSA) and *L. pneumophila* with ≥99% growth reduction at 2–16 times the MIC [14]. It also showed dose-related reduction of established biofilms and inhibited *E. coli* biofilm propagation at sub-MIC concentration [14].

## 7.8 Microbiologic Activity of Omadacycline

OMAD microbial activity is similar to other tetracyclines, with broad-spectrum against aerobic and anaerobic gram-positive and gram-negative bacteria and atypical organisms. Similar to ERV, it is active against MDR-resistant strains, including tetracycline resistance. The susceptibility breakpoints for OMA are *Enterobacteriaceae* ≤ 4.0 μg/ml, *S. aureus* ≤0.5 μg/ml, *E. faecalis*, and *Streptococcus* spp. ≤0.25 μg/ml [13].

Against gram-positive bacteria, OMAD antimicrobial activity appears very similar to ERV but there is no direct comparison between the t*wo agents for most species* or strains tested in parallel. Against *S. aureus* including MRSA and MDR strains, the $MIC_{90}$ was 0.5 μg/ml; all streptococcal strains were inhibited by 0.5 μg/ml; and both *E. faecalis and E. faecium* $MIC_{90}$ *(including VRE) were 0.5 μg/ml* [13]. However, in a recent study of 80 isolates of VRE, of the *54 E. faecium* isolates, 96.3% were susceptible to tigecycline, 96.3% to ERV, and 77.8% to OMAD; and of the 26 *E. faecalis* isolates, 100% were susceptible to tigecycline, 96.15% to ERV, and only 7.6% [2] to OMAD [16].

OMAD is highly active against respiratory gram-negative bacteria with 99% of *H. influenzae* inhibited by ≤2 μg/ml and 100% of *M. catarrhalis* by 1 μg/ml [13]. Against *Enterobacteriaceae, E. coli* $MIC_{90}$ was 2 μg/ml and *Klebsiella* spp. $MIC_{90}$ was 4 μg/ml [13]. Against MDR gram-negative bacteria, 91.5–95.5% of *Acinetobacter* spp. were inhibited by ≤4 μg/ml, and 82.2% of *S. maltophilia* were also inhibited by OMAD. For other MDR-*Enterobacteriaceae,* OMAD inhibited 85.3% of ceftazidime-resistant strains and 52.7% of imipenem-nonsusceptible isolates [17]. However, OMAD is not active against *Pseudomonas, Proteus, Morganella,* or *Providentia* species [Medical Letter on drugs and therapeutics. JAMA 2019; 32: 457–8].

The anaerobic activity of OMAD was similar to that of tigecycline with $MIC_{90}$ values outlined: *B. fragilis* 4 μg/ml, *Prevotella* spp. 2 μg/ml, *Bacteroides vulgatus* 1 μg/ml, *Bacteroides ovatus* 8 μg/ml, *C. difficile 0.5 μg/ml, C. perfringens* 16 μg/ml, and anaerobic gram-positive cocci—1 μg/ml [18]. It is highly active against atypical bacteria, similar to doxycycline and azithromycin, with $MIC_{90}$ 0.06 μg/ml *for M. pneumoniae*, *C. pneumoniae*, and *L. pneumophila* [14]. OMAD is also very active against rapidly growing mycobacteria with $MIC_{90}$ of 2 μg/ml for *M. abscessus*, 0.5 μg/ml for *Mycobacterium fortuitum*, and 0.25 μg/ml for *Mycobacterium chelonae* [7, 19]. See Tables 7.1 and 7.2 for summary of the in vitro activity.

### 7.8.1 Resistance to Omadacycline

The section on resistance to ERV (7.4) also applies to OMAD. As of 2020, no strains of bacteria with induced resistance have been reported [13]. Most studies on antibiotic resistance general focused on stable genetic mutations or acquisition of resistant genes which result in resistance to all cells within bacterial population. However, heteroresistance in apparently susceptible strain is being increasingly recognized as playing a significant role in antimicrobial resistance [20]. Clinical laboratory susceptibility testing may incorrectly report an isolate as susceptible even when there is minor resistant subpopulation. Subsequent antibiotic therapy selects for the resistant subpopulation which eventually leads to treatment failure. A recent study from China showed that OMAD heteroresistance in a sample of 263 *S. aureus* occurred in 3.17% of MRSA and 12.78% of MSSA [21]. The heteroresistance was more frequent in isolates with MICs ≥0.5 μg/ml. Similar heteroresistance was previously reported with ERV as well. Molecular studies indicated that the heteroresistance were related to overexpression or mutation of genes encoding efflux pump proteins.

It was thought until recently that tetracycline resistance in human pathogens occurred almost exclusively by ribosomal protection and antibiotic efflux. However, there is evidence bacteria carrying tetracycline inactivating enzymes are present not only in the environment but in human commensals and pathogenic gram-negative bacteria. Tet(X) was the only known enzyme capable of inactivating tetracycline and tigecycline found in nonpathogenic bacteria, but the emergence of Tet (X), Tet (x3), Tet(x4), and Tet(x7) can be found in clinical pathogens and are capable of inactivating all tetracyclines including ERV and OMAD [9]. The widespread use of these modern tetracyclines will promote further dissemination of these tetracycline inactivating genes through plasmids [9].

## 7.9 Clinical Efficacy and Safety of Omadacycline

OMAD has been studied in three phase 3 trials before FDA approval: the OASIS 1 and 2 trials for ABSSSIs and the OPTIC trial for CAP [13]. Prior to the phase 3 trials, a RCT phase 2 study was done in patients with ABSSSIs comparing OMAD and linezolid (both initially IV with option to transition to oral) primarily to assess safety and tolerability and secondarily to assess clinical response, which was not significantly different between the treatments.

The OASIS-1 and OASIS-2 are multicenter, double-blind, noninferiority RCTs for ABSSSIs. The OASIS-1 study randomized 627 patients to IV OMAD 100 mg twice daily for 2 doses then 100 mg daily or linezolid 600 mg twice daily with the option of switching to oral OMAD 300 mg daily or oral linezolid 600 mg twice daily for 7–14 days [22]. The primary endpoint was early clinical response at 48–72 h, ≥20% reduction in lesion size and clinical success rate at 7–14 days after treatment. OMAD was noninferior to linezolid with clinical success after treatment of 86.1% and 83.6%, respectively.

The OASIS-2 study was of similar design, but 720 patients were randomized to oral doses only: OMAD 450 mg daily for 2 doses and then 300 mg daily or linezolid 600 mg twice daily for 7–14 days. OMAD again demonstrated noninferiority at early response and clinical success at the post-treatment assessment, 84.2% vs 80.8% for linezolid [23]. The 2 trials included patients with wound infections, cellulitis, and major abscess. *S. aureus* was detected in 74.7% with MRSA in 32.4% and gram-negative aerobes in 10.3% and gram-negative anaerobes in 5.2% [24]. Interestingly, the clinical response between the two treatment arms was similar even for patients with mixed infections with gram-negative bacteria (aerobes or anaerobes), despite linezolid lacking activity against these bacteria. This would suggest that in skin and soft tissue infection with mixed gram-positive and gram-negative bacteria, covering the gram-negative pathogens may not be necessary.

The Omadacycline for Pneumonia Treatment in the Community (OPTIC) trial was a phase 3, double-blind, multicenter, noninferiority RCT comparing OMAD to moxifloxacin [25]. It compared OMAD 100 mg IV daily (initial 2 doses every 12 h), with an option to transition to 300 mg orally daily after 3 days, or moxifloxacin 400 mg IV daily, with an option to transition to oral 400 mg daily after 3 days for a total of 7–14 days. The primary endpoints were early clinical response at 72–120 h and clinical success 5–10 days after treatment. OMAD was noninferior to moxifloxacin with similar clinical success, 87.6% vs 85.1%, respectively, at the post-treatment evaluation.

The safety of OMAD is best assessed from the pooled data of the phase 3 trials. As with most tetracyclines transient, nausea and vomiting and mild elevations of liver transaminases were the most common adverse effects of OMAD in phase 3 studies [26]. Drug-related adverse events were reported in 22.0% with OMAD compared to 16.1% for linezolid and 17.8% for moxifloxacin, but serious drug-related adverse events were rare in all treatment arms, 0.2%, 0.1%, and 0.5%, respectively [26]. GI-related side effects were the most common events in all treatment groups. Diarrhea occurred less frequently with OMAD (2.4%) and linezolid (2.9%) compared to moxifloxacin (8%). There was no *C. difficile* colitis in patients treated with OMAD and linezolid but in 8 (2.1%) patients treated with moxifloxacin, and this has been noted for the tetracycline class [26]. The newer agents of the tetracycline class, as with older agents, should be avoided in pregnancy, infancy, and childhood up to 8 years of age due to risk of tooth discoloration, enamel hypoplasia, and inhibition of bone growth (see package insert).

## 7.10 Place in Therapy with the Modern Tetracyclines

ERV has been approved for only cIAIs and was less effective than levofloxacin for cUTIs; thus it will have a limited role in the management of infections. It is most suitable for hospital-associated IAIs where there is greater risk of MDR *Enterobacteriales,* including carbapenem- and fluoroquinolone-resistant strains and for the rare cases with MRSA or VRE mixed infections. However, the phase 3 clinical trials did not include adequate number of patients with these difficult to treat organisms to prove its efficacy, and it was not superior to the comparator agent. Post-marketing prospective observational studies will be needed to show the clinical response against carbapenem-resistant *Enterobacteriaceae.*

There are numerous agents now available for treatment of acute bacterial skin and soft tissue infections, and in the past 5–6 years, four new drugs with coverage for MRSA infections have been approved for this indication (tedizolid, oritavancin, dalbavancin, and OMAD). Furthermore, there are other pre-existing agents including linezolid, tigecycline, vancomycin, daptomycin, cotrimoxazole, and doxycycline. OMAD has a narrow advantage over some of these agents such as its availability in parenteral and oral formulation and its wider spectrum to cover gram-negative bacteria. Thus it may be more suitable for treating mixed bacterial infections (i.e., post-surgical wound infections, deep diabetic foot infections with chronic ulcers, severe deep sacral ulcer infections). However, its superiority over current standard agents is yet to be proven.

OMAD is more suitable for hospital-acquired pneumonia (including ventilator-associated pneumonia) where MDR-gram negative and MRSA are more common. Thus, RCTs are needed for these indications but post-marketing prospective cohort studies on carbapenem-resistant gram-negative pneumonia with or without MRSA would be useful. Although OMAD would cover all the possible bacteria causing CAP, its use in this indication would represent overkill, and there are several alternatives that would be less expensive. Moreover, there is no great need for new treatment of CAP, and use of OMAD for common infections will encourage widespread resistance to the new class of tetracyclines.

## References

1. Nelson Ml, Levy SB. The history of the tetracyclines. Ann N Y Acad Sci. 2011;1241:17–32.
2. Chopra I, Roberts M. Tetracycline antibiotics: mode of action, applications, molecular biology, and epidemiology of bacterial resistance. Microbiol Mol Biol Rev. 2001;65:232–60.
3. Alosaimy S, Abdul-Mutabbir JC, Kebriaei R, Jorgensen SCJ, Rybak MJ. Evaluation of eravacycline: a novel fluorocycline. Pharmacotherapy. 2020;40:221–38.
4. Morissy I, Olesky M, Hawser S, et al. *In vitro* activity of eravacycline against gram-negative bacilli isolated in clinical laboratories worldwide from 2013 to 2017. J Clin Microbiol. 2020:64; https://journals.asm.org/doi/10.1128/aac.01699-19?permanently=true.
5. Lee YR, Burton CE. Eravacycline, a newly approved fluorocycline. Eur J Clin Microbiol Infect Dis. 2019;38:1787–94.
6. Sutcliffe JA, O'Brien W, Fyfe C, Grossman TH. Antibacterial activity of eravacycline [TP-434], a novel fluorocycline, against hospital and community pathogens. Antimicrob Agents Chemother. 2013;57:5548–58.
7. Kaushik A, Ammerman NC, Martins O, Parrish NM, Nuermberger EL. *In vitro* activity of new tetracycline analogs omadacycline and eravacycline against drug-resistant clinical isolates of *Mycobacterium abscessus*. Antimicrob Agents Chemother. 2019;63:e00470–19.
8. Zheng JX, Zhi-Wei L, Sun X, et al. Overexpression of OqxAB and MacAB efflux pumps contributes to eravacycline resistance and heteroresistance in clinical isolates of *Klebsiella pneumoniae*. Emerg Microbes Infect. 2018;7:1–11.
9. Gasparrini AJ, Markely JL, Kumar H, et al. Tetracycline-inactivating enzymes from environmental, human commensal, and pathogenic bacteria cause broad-spectrum tetracycline resistance. Commun Biol. 2020;3:241. https://www.nature.com/articles/s42003-020-0966-5
10. Solomkin J, Evans D, Slepavicius A, et al. Assessing the efficacy and safety of eravacycline vs ertapenem in complicated intra-abdominal infections in the investigating gram-negative infections treated with eravacycline [IGNITE 1] trial: a randomized clinical trial. JAMA Surg. 2017;152:224–32.
11. Solomkin JS, Gardovskis J, Lawrence K, et al. IGNITE4: results of a phase 3, randomized, multicenter, prospective trial of eravacycline vs meropenem in the treatment of complicated intra-abdominal infections. Clin Infect Dis. 2019;69:921–9.
12. Eljaaly K, Ortwine JK, Shaikhomer M, Almangour TA, Bassetti M. Efficacy and safety of eravacycline: a meta-analysis. J Global Antimicrob Resistance. 2021;24:424–8.
13. Barber KE, Bell AM, Wingler MJB, Wagner JL, Stover KR. Omadacycline enters the ring: a new antimicrobial contender. Pharmacotherapy J Human Pharm Drug Ther. 2020;38:1194–204.
14. Karlowsky JA, Steenbergen J, Zhanel GG. Microbiology and preclinical review of omadacycline. Clin Infect Dis. 2019;69(Suppl 1):S6–15.
15. Agwuh KN, MacGowan A. Pharmacokinetics and pharmacodynamics of the tetracyclines including glycyclines. J Antimicrob Chemother. 2006;58:256–65.
16. Chua MFP, Nida SS, Lawhorn J, Sundareshan V. A comparative analysis of *in vitro* susceptibilities of vancomycin-resistant enterococci against doxycycline, minocycline, tigecycline, eravacycline and omadacycline. Open For Infect Dis. 2020;7(Suppl 1):S790. https://doi.org/10.1093/ofid/ofaa439.1766.
17. Pfaller MA, Huband MD, Shortridge D, Flamm RK. Surveillance of omadacycline activity against clinical isolates from the United States and Europe as part of the 2016 SENTRY antimicrobial program. Antimicrob Agents Chemother. 2018;62:1–7.

18. Stapert L, Wolfe C, Shinabarger D, Marra A, Pillar C. *In vitro* activities of omadacycline and comparators against anaerobic bacteria. Antimicrob Agents Chemother. 2018;62:e00047–18.
19. Shoen C, Bernard D, Sklaney M, Cyanamon M. *In vitro* activities of omadacycline against rapidly growing mycobacteria. Antimicrob Agents Chemother. 2019;63:e02522–18.
20. Band VI, Weiss DS. Heteroresistance: a cause of unexplained antibiotic treatment failure? PLoS Pathog. 2019;15:e1007726.
21. Bai B, Pu Z, Xu G, et al. *In vitro* activity and heteroresistance of omadacycline against clinical *Staphylococcus aureus* isolates from China reveal the impact of omadacycline susceptibility by branched-chain amino acid transport system II carrier protein, Na/Pi cotransporter family protein, and fibronectin-binding protein. Fron Micobiol. 2019; https://doi.org/10.3389/fmicb.2019.02546.
22. O'Riordan W, Green S, Overcash JS, et al. Omadacycline for acute bacterial skin and skin-structure infections. N Engl J Med. 2019;380:528–38.
23. O'Riordan W, Cardenas C, Shin E, et al. Once daily oral omadacycline versus twice-daily linezolid for acute bacterial skin and skin structure infections [OASIS-2]: a phase 3, double-blind, multicenter, randomized, controlled, non-inferiority trial. Lancet Infect Dis. 2019;19:1080. https://doi.org/10.1016//s1473-3099[19]30275-0.
24. Abrahamian FM, Sakoulas G, Tzanis E, Manley A, Steenbergen J, Das AF, Eckburg PB, McGovern PC. Omadacycline for acute bacterial skin and skin structure infections. Clin Infect Dis. 2019;69(Suppl 1):S23–32.
25. Stets R, Popescu M, Gonong JR, et al. Omadacycline for community-acquired bacterial pneumonia. N Engl J Med. 2019;380:517–27.
26. Opal S, File TM Jr, van der Poll T, Tzanis E, Chiura S, McGovern PC. An integrated safety summary of omadacycline, a novel aminomethylcycline antibiotic. Clin Infect Dis. 2019;69(Suppl 1):S40–7.

# Chapter 8
# Pleuromutilin: A New Class of Antibiotic: Lefamulin

## 8.1 Introduction

Pleuromutilin was discovered in the 1950s, isolated from an edible mushroom, *Pleurotus mutilus* (renamed *Clitopilus scyphoides*), and semisynthetic derivatives were developed in 1979 and 1999 for use in veterinary medicine [1]. Despite use of tiamulin and valnemulin in veterinary medicine for over 30 years, bacterial resistance was uncommonly reported [1]. Valnemulin (marketed as Econor), used for treating swine with dysentery, ileitis, colitis, and pneumonia, has been used on compassionate basis to treat three patients with primary immunodeficiency and refractory mycoplasma infections (recovered from joints or cerebrospinal fluid). After failing therapy with doxycycline, quinolone, and macrolides, the patients were treated successfully with the pleuromutilin antibiotic [2]. Retapamulin, a topical agent, was the first pleuromutilin approved for human use for infected small lacerations or abrasions (United States [US] Food and Drug Administration [FDA] 2007). In 2019, lefamulin (marketed as Xenleta) was the first pleuromutilin approved by the FDA for systemic use in the treatment of community-acquired bacterial pneumonia (CABP).

## 8.2 Chemical Structure and Mechanism of Action

Lefamulin is a semisynthetic pleuromutilin with a tricyclic mutilin core that is essential for antimicrobial activity and a C14 side chain which provides the main pharmacodynamics and antimicrobial functions [3]. Modifications of the C14 side chain results in improved solubility and metabolic stability with enhanced antimicrobial activity but also allow lefamulin to overcome bacterial mutations and resistance [3]. Lefamulin inhibits protein synthesis by preventing the binding of tRNA



I. W. Fong, *New Antimicrobials: For the Present and the Future*, Emerging Infectious Diseases of the 21st Century,
https://doi.org/10.1007/978-3-031-26078-0_8

for peptide transfer, similar to the oxazolidinones, with tight binding of the drug to the target site [3]. This antibiotic has bacteriostatic activity against some bacteria, but bactericidal against others including *Mycoplasma pneumoniae* [4].

## 8.3 Antimicrobial Activity

Lefamulin is active against all aerobic gram-positive bacteria except *Enterococcus faecalis* and limited gram-negative bacterial activity [3]. It has potent bactericidal activity against most aerobic gram-positive bacteria: *Staphylococcus aureus* including methicillin-resistant strains (MRSA), vancomycin-intermediate strains (VISA), and vancomycin-resistant strains (VRSA), coagulase-negative staphylococcus (CNS), streptococcus spp. including multidrug-resistant *Streptococcus pneumoniae*, β-hemolytic and alpha-hemolytic streptococci, and *Enterococcus faecium* including vancomycin-resistant strains (VRE), but not *E. faecalis* [5].

It has limited gram-negative bacterial activity, mainly against respiratory pathogens, *Haemophilus influenzae* and *Moraxella catarrhalis*, and genital pathogens such as multidrug-resistant *Neisseria gonorrhoeae*, *Chlamydia trachomatis*, and *Mycoplasma genitalium* [3, 5, 6]. Lefamulin has no activity against *Enterobacterales* and *Pseudomonas* spp. but is active against atypical pathogens, *M. pneumoniae*, *Chlamydia pneumoniae*, and *Legionella pneumophila* [3]. It has activity against some anaerobes, including *Clostridium perfringens*, *Cutibacterium acnes*, *Fusobacterium* spp., *Peptostreptococcus* spp., *Prevotella* spp., and *Porphyromonas* spp., but not *Clostridioides difficile* or *Bacteroides fragilis* [3]. Table 8.1 summarizes the in vitro activity of lefamulin.

## 8.4 Microbial Resistance to Lefamulin

The development of resistance to lefamulin is expected to be rare as spontaneous mutation frequencies are low ($\leq 10^{-9}$), slow stepwise resistance development at sub-MIC in vitro [1]. The mechanisms of resistance to other pleuromutilins used in veterinary medicine are frequently the result of ribosomal protein point mutations of the 23S RNA that affect the peptidyl transferase structure. Resistance to lefamulin has been documented in vitro with alteration of the target site [1]. The mechanisms of resistance include mutations in the 23S rRNA and *rplC* and *rplD* genes that encode for ribosomal proteins L3 and L4. High level resistance to lefamulin in *Mycoplasma* spp. can result from a single mutation in the 23S rRNA, and *rplC* mutation in *S. aureus* can result in resistance [1]. Other mechanisms of resistance in *S. aureus* include mutations in the *cfr* gene and transposon and plasmid that encode for the ABC-F transporter that mediates resistance through target site alterations

**Table 8.1** Comparative in vitro activity of lefamulin

| Microorganisms | Lefamulin | Vancomycin | Azithromycin | Moxifloxacin |
|---|---|---|---|---|
| | $MIC_{50}/MIC_{90}$ | $MIC_{50}/MIC_{90}$ | $MIC_{50}/MIC_{90}$ | $MIC_{50}/MIC_{90}$ |
| *Gram-positive* | *μg/ml* | *μg/ml* | *μg/ml* | *μg/ml* |
| *S. aureus* (*n* = 5527) | 0.12/0.12 | 1/1 | — | — |
| MRSA | 0.12/0.25 | 1/1 | — | — |
| *S. Pyogenes* (*n* = 267) | 0.03/0.03 | 0.25/0.5 | — | — |
| *S. agalactiae* (*n* = 334) | 0.03/0.03 | 0.5/0.5 | — | — |
| *S. viridans* gp. (*n* = 245) | 0.12/0. 5 | 0.5/0.5 | — | — |
| *S. pneumoniae* (*n* = 1473) | 0.12/0.25 | 0.25/0.5 | — | — |
| *E. Faecium* (536) | 0.12/4 | >16/>16 | — | — |
| *Gram-negative* | | | | |
| *H. Influenzae* (*n* = 360) | 1/2 | — | 1/2 | ≤0.5/≤0.5 |
| *M. Catarrhalis* (*n* = 253) | 0.12/0.25 | — | ≤0.25/≤0.25 | ≤0.5/≤0.5 |
| *Atypical organisms* | | | | |
| *L. Pneumophila* (*n* = 30) | 0.12/0.5 | — | 0.06/0.12 | 0.06/0.12 |
| *C. pneumoniae* (*n* = 50) | 0.02/0.04 | — | — | — |
| *M. pneumoniae* (*n* = 50) | 0.006/0.006 | — | — | — |
| [a]*STD pathogens* | | | | |
| *N. Gonorrhoeae* (*n* = 251) | 0.25/1 | — | — | — |
| *M. Genitalium* (26) | 0.016–0.063 | — | — | — |

Adapted from Ref. [3]
Abbreviations: *MRSA* methicillin-resistant *S. aureus*, *STD* sexually transmitted disease pathogens
[a]Refs. [5, 14, 15]

and efflux pumps [1, 7]. However, *S. aureus* resistance to lefamulin was rare, 0.18% of 5527 isolates [7]. Resistance among CNS is higher at 3.4% with mutations occurring in the *vga (A)* gene, *rplD* gene, and the *cfr* gene [7].

## 8.5 Pharmacokinetics

Lefamulin is available in parenteral and oral formulation and the standard dose is 150 mg intravenously (IV) over I h every 12 h and 600 mg orally every 12 h (see package insert of lefamulin [Xenleta]). After IV dosing of 150 mg the maximum serum concentration ($C_{max}$) varies from 1.90 μg/ml (single dose) to 2.06 μg/ml (multiple dosing) with area under the concentration curve ($AUC_{0-24}$) over 24 h of 14.1–16.5 μg.h/ml [8]. About 25% of the oral dose is absorbed but 600 mg oral dosing provided equivalent blood levels as the 150 mg IV dosing [8]. The terminal half-life was 9–12 h and the protein binding 80–87% [3]. The drug is widely distributed and concentrates in the lung epithelial lining fluid and macrophages [9].

Metabolism of lefamulin occurs by the CYP450 enzymes, as a substrate and inhibitor of CYP3A and 77.3–88.5% is excreted in the feces and 13% in the urine [3]. There is no dosage adjustment necessary for patients with renal dysfunction, including those on dialysis, and lefamulin is not recommended for those with moderate-severe hepatic impairment [10].

Based on animal studies and population pharmacokinetic models, the area under the curve (AUC)/MIC ratio is the best parameter to predict lefamulin efficacy or activity [3, 10]. Present IV and oral dosing recommendations are predicted to produce 1–2-$\log_{10}$ colony-forming units (CFU) reductions of *S. aureus* and S. *pneumoniae* in about 92–99% of the time [3].

## 8.6 Clinical Efficacy

Two phase 3 multinational, randomized, double-blind clinical trials for CABP, LEAP 1, and LEAP 2 [11, 12], and one phase 2 trial for skin and skin structure infections [13] were performed to assess the clinical efficacy of lefamulin.

In LEAP 1, 551 patients were randomized to IV lefamulin 150 mg every 12 h for 5–7 days or IV moxifloxacin 400 mg every 24 h for 7 days. Patients could switch to oral therapy (lefamulin 600 mg every 12 h or moxifloxacin 400 mg every 24 h) after 3 days if predefined improvement criteria occurred. Patients with MRSA infection received 10 days of treatment and the moxifloxacin-treated group was allowed linezolid. In LEAP 2, 738 patients were randomized to oral lefamulin 600 mg every 12 h for 5 days or oral moxifloxacin 400 mg every 24 h for 7 days. In both trials, lefamulin was noninferior to moxifloxacin for the primary early clinical response at 96 ± 24 h after initiation of therapy and the post-therapy test of cure, and this was maintained in the pooled analysis, 89.3 vs 90.5% early clinical response [10]. The most common pathogens were *S. pneumoniae, S. aureus (MSSA), H. influenzae,* and atypical pathogens (*M. pneumoniae*, *L. pneumophila*, *C. pneumoniae*).

In a phase 2 clinical trial of skin and skin structure infections, patients were randomized to two different dosing of IV lefamulin (100 mg or 150 mg every 12 h) or IV vancomycin 1 g every 12 h for 5–14 days [11]. Of the 210 randomized patients, 186 (88.6%) completed the study and the primary endpoint was clinical success at the test-of-cure visit 7–14 days after treatment. The response rates were similar among the three groups, 90.0% for lefamulin 100 mg daily, 88.9% for lefamulin 150 mg daily, and 92.2% in the vancomycin group. Of the patients with isolated pathogens, *S. aureus* was the commonest isolate in all groups (81.8%–87.2%) and MRSA was responsible in similar rates among the three groups, 85.3% and 87.5% in the lefamulin groups, and 82.1% in the vancomycin group. Microbial success at test of cure was also similar between the groups (80% and 84.3%) for the lefamulin groups and 82.4% for the vancomycin group.

## 8.7 Adverse Effects

Overall, lefamulin was generally well tolerated and adverse events were mild to moderate in most cases. The most common adverse events with IV dosing were infusion reaction (2.2–5.7%) and gastrointestinal (GI) symptoms, and with oral dosing the GI adverse events were lower when lefamulin was administered after food [8]. In comparison to vancomycin, IV lefamulin was associated with higher rate of infusion site phlebitis (2.8–5.7%), vulvovaginal fungal infection, tinnitus, and increased creatine phosphokinase than vancomycin, but vancomycin was associated with greater pruritus, nausea, and liver enzyme elevation [13]. In the LEAP 1 trial, lefamulin was associated with mild increase in QT interval ($n = 3$) but less than with moxifloxacin ($n = 5$). In the LEAP I trial IV lefamulin was associated with less risk of diarrhea (0.7%) than IV moxifloxacin (7.7%), but this was reversed in the LEAP-2 trial with oral dosing, 12.2% vs 1.1%, respectively. No patients had *C. difficile* colitis.

## 8.8 Place in Therapy

Although lefamulin is the first new class of antibiotic introduced for treatment of CABP in 15 years, it is the second new antibiotic approved for treatment of CABP in North America over these years, the first being omadacycline, a new tetracycline. Both agents are available for IV and oral therapy and have potent antimicrobial activity against the bacteria causing CABP, including atypical organisms. Because of the potent activity of omadacycline against gram-negative pathogens, it is more suitable for hospital-acquired pneumonia. The cost of lefamulin will limit its use for CABP compared to currently available agents, the wholesale cost in the US per day is $205 for the IV preparation and $275 for the tablets. It may be used for situations requiring quinolone- and macrolide-sparing therapy or where MRSA infection is a concern.

Lefamulin is active against sexually transmitted pathogens and may have a role in the treatment of drug resistant strains of *N. gonorrhoeae* and *M. genitalium,* as resistance to ceftriaxone, tetracycline, macrolides, and quinolone does not usually result in cross-resistance. However, clinical studies are needed in this area, particularly for multidrug-resistant infections.

## References

1. Paukner S, Riedl R. Pleuromutilins: potent drugs for resistant bugs—mode of action and resistance. Cold Spring Harb Perspect Med. 2017;7:a027110.
2. Heilmann C, Jensen L, Lundstrom K, Windsor D, Webster. Treatment of resistant mycoplasmas infection in immunocompromised patients with a new pleuromutilin antibiotic. J Infect. 2001;43:234–8.

3. Veve MP, Wagner JL. Lefamulin: review of a promising novel pleuromutilin antibiotic. Pharmacotherapy. 2018;38:935–46.
4. Waiters KB, Crabb DM, Duffy LB, et al. In vitro activities of lefamulin and other antimicrobial agents against macrolide-susceptible and macrolide-resistant *mycoplasma pneumoniae* from the United States, Europe and China. Antimicrob Agents Chemother. 2017;61:e02008–16.
5. Paukner S, Gruss A, Jensen JS. In vitro activity of lefamulin against sexually transmitted bacterial pathogens. Antimicrob Agents Chemother. 2018;62:e02330–17.
6. Jacobson S, Paukner S, Golparian, et al. In vitro activity of the novel pleuromutilin lefamulin [BC-3781] and effect of efflux pump inactivation on multidrug-resistant and extensively drug-resistant *Neisseria gonorrhoeae*. Antimicrob Agents Chemother. 2017;61:e01497–17.
7. Paukner S, Sader HS, Ivezic-Schoenfeld Z, et al. Antimicrobial activity of the pleuromutilin antibiotic BC-3781 against bacterial pathogens isolated in the SENTRY antimicrobial surveillance program in 20210. Antimicrob Agents Chemother. 2013;57:4489–95.
8. Wicha WW, Prince WT, Lell C, Heilmayer W, Gelone SP. Pharmacokinetics and tolerability of lefamulin following intravenous and oral dosing. J Antimicrob Chemother. 2019;74(Suppl 3):iii9–26.
9. Zeitlinger M, Schwarmeis R, Burian A, et al. Simultaneous assessment of the pharmacokinetics of a pleuromutilin, lefamulin, in plasma, soft tissues, and pulmonary epithelial lining fluid. J Antimicrob Agents Chemother. 2016;71:1022–6.
10. Watkins RR, File TM Jr. Lefamulin: a novel semisynthetic pleuromutilin antibiotic for community-acquired bacterial pneumonia. Clin Infect Dis. 2020;71:2757–62.
11. File TM, Goldberg L, Das A, et al. Efficacy and safety of intravenous-to-oral lefamulin, a pleuromutilin antibiotic, for the treatment of community-acquired bacterial pneumonia: the phase III Lefamulin evaluation against pneumonia [LEAP I] trial. Clin Infect Dis. 2019;69:1856–67.
12. Alexander E, Goldberg L, Das AF, et al. Oral lefamulin vs moxifloxacin for early clinical response among adults with community-acquired bacterial pneumonia: the LEAP 2 randomized clinical trial. JAMA. 2019;322:1661–71.
13. Prince WT, Ivezic-Schoenfeld Z, Lell C, Tack KJ, Novak R, Obermayr F, Talbot GH. Phase II clinical study of BC-3781, a pleuromutilin antibiotic. In treatment of patients with acute bacterial skin and skin structure infections. Antimicrob Agents Chemother. 2013;57:2087–94.
14. Jacobsson S, Paukner S, golparian D, et al. In vitro activity of the novel pleuromutilin lefamulin [BC-3781] and efflux pump inactivation on multidrug-resistant and extensively drug-resistant *Neisseria gonorrhoeae*. Antimicrob Agents Chemother. 2017;61:e01497–17.
15. Jensen JS, Paulner S. Lefamulin is highly active in vitro against multi-drug resistant *Mycoplasma genitalium* strains. Poster/abstract. In: 27th European congress of clinical microbiology and infectious diseases, 22–25 April. Vienna; 2017.

# Chapter 9
# New Anti-tuberculous Drugs: Bedaquiline, Delamanid, and Pretomanid

## 9.1 Introduction

Tuberculosis (TB) caused by *Mycobacterium tuberculosis* is the leading cause of death globally from a single infectious disease [1], and there is no visible end in sight despite "The End Tuberculosis Strategy" initiated by the World Health Organization (WHO). In 2018, it was estimated that ten million persons had incident TB with 1.5 million related deaths [Center of Disease Control and Prevention [CDC], Weekly/March 20, 2020/69; 281-5]. Multidrug-resistant (MDR)-TB occurred in 3.4% of new cases and 18% of cases previously treated, estimated to be 465,000 cases in 2019. Hence, there is a great need for tolerable, new agents to combat MDR-TB.

Prior to 2010, there was no new drug class introduced for the treatment of TB and the only new drug approved was rifapentine, a long acting rifamycin, which was approved by the FDA in 1998. Since then rifapentine has been rarely used in clinics, but there is evidence that in combination with isoniazid it can shorten treatment of latent TB to 1 month [2], and 4-month rifapentine-based regimen containing moxifloxacin was noninferior to standard 6-month regimen [3]. However, rifapentine is not effective against MDR-TB. In the last decade, three new drugs (bedaquiline, delamanid, and pretomanid) were introduced on the global scene for treatment of MDR-TB.

## 9.2 Bedaquiline

Bedaquiline (marketed as Sirturo) was approved by the FDA in December 2012 for the treatment of MDR-TB. Bedaquiline (BDQ) is the first drug in a new class approved for treatment of TB since approval of rifampin in 1971 in the US [4]. It belongs to the



I. W. Fong, *New Antimicrobials: For the Present and the Future*, Emerging Infectious Diseases of the 21st Century,
https://doi.org/10.1007/978-3-031-26078-0_9

diarylquinoline class of agents which are closely related to the fluoroquinolones but with different modes of action. BDQ contains a central heterocyclic nucleus with alcohol and amine side chains which are responsible for the anti-TB activity [5]. It is the only anti-TB drug that targets adenosine triphosphate (ATP) by inhibiting the mycobacterial ATP synthase (essential for cell function) without interaction with human ATP synthase [6].

### 9.2.1 Microbial Activity of Bedaquiline

BDQ is a narrow spectrum antibiotic with antimicrobial activity limited to mycobacteria. It has potent activity against *M. tuberculosis* and most nontuberculous mycobacteria (NTM), rapid or slow growing, but *Mycobacterium xenopi* is naturally resistant [4]. BDQ has bactericidal activity against replicating or dormant mycobacteria, intracellular or extracellular forms. Compared to isoniazid (INH) and rifampin, BDQ has greater potency exceeding the bactericidal activities of both by 1 log unit [6], with mean MIC of 0.03 μg/ml against MDR—*M. tuberculosis* [7].

### 9.2.2 Resistance to Bedaquiline

Since the expedited approval of BDQ for MDR-TB in 2012, >2500 patients received the drug by 2015, and by the end of 2017, 68 countries began using the drug [4]. However, resistance to BDQ has been reported soon after its introduction from acquired resistance [8], and primary resistance in treatment-naïve patients has recently been reported [9]. In vitro resistance to BDQ was previously shown to be due to target-based resistance and is mainly related to chromosomal mutations. Inadequate and incomplete treatment can lead to selection of resistant mutants (WHO guidelines). The three main resistance mechanisms are as follows: (i) mutations within the *atpE* gene, target-based mutations which causes high increase in MIC (10–128 times); (ii) non-target based mutations in the *Rvo678* gene, regulating the expression of the Mmp55-mmpl5 efflux pumps, causes low level resistance (two-eightfold increase in MIC) and cross-resistance to clofazimine (CFZ); and (iii) non-target mutation in *pepQ* gene may be related to higher antibiotic efflux with cross-resistance to CFZ [4, 10].

Presently there is no universal definition of BDQ resistance and no standardized drug testing protocol. The European Committee on Antimicrobial Susceptibility Testing (EUCAST) guidelines, MIC breakpoints for BDQ are susceptible <0.25 μg/ml and resistant, ≥0.25 μg/ml [4]. Resistance to BDQ can be induced in laboratory experiments at 4 times the MIC with resistant mutants at $5 \times 10^{-7}$ and at 8 times the MIC with a frequency of $5 \times 10^{-8}$ [6].

BDQ resistance has also been demonstrated in clinical studies where drug susceptibility testing (DST) to BDQ were performed at baseline and 24 weeks or later. In study C208, 13 of 28 patients had four-fold increase in MIC and in study C209, 12 of 24 patients also demonstrated >4-fold increase in MIC attributed to *Rvo678* mutations [10]. Clinical use of BDQ in patients with MDR-TB and extensive drug-resistant (XDR) have resulted in resistance to BDQ and cross-resistance to CFZ and rapid development of resistance to delamanid [10, 11]. Low level resistance at baseline without prior exposure to BDQ or CFZ of 1.9% to 2.3% due to *Rv678* variants is also of concern [12].

In a recent study from China, 277 patients with MDR-TB received BDQ with 6 of 277 (2.2%) showing BDQ resistance at baseline [13]. Of 94 cases that had serial DST, 11 (11.7%) acquired reduced susceptibility to BDQ and 3 of 5 cases with acquired resistance failed to culture-convert. Six of 11 BDQ-resistant isolates had *Rv0678* mutations, but no mutations were detected in 3 other BDQ resistance-associated genes. Acquired BDQ resistance was associated with greater risk of treatment failure compared to baseline BDQ resistance.

## 9.3 Pharmacokinetics of Bedaquiline

BDQ is well absorbed after oral administration, and peak plasma concentration ($C_{max}$) is achieved in 4–6 h, and the recommended dose is 400 mg once daily in patients with MDR-TB [14]. Administration of the drug with food increases the oral bioavailability, and it should be taken with food. BDQ is highly protein bound (≥99.7%), yet in mice it is widely distributed in tissues, including the lungs and spleen but brain uptake was low [14]. BDQ and its active metabolite penetrate freely in the cerebrospinal fluid (CSF) of patients with pulmonary TB and no evidence of meningeal inflammation [15].

The mean plasma half-life of BDQ is 24–30 h but the terminal half-life is extremely long (4–5 months) [4] and the drug is primarily metabolized in the liver by cytochrome P450 (CYP) isoenzyme 3A4 into an active N-monodes-methyl metabolite (M2), which is fivefold less active than BDQ [14]. BDQ is mainly excreted in the feces (75–85% of unchanged drug eliminated in 24 h) and urinary excretion is negligible. The long terminal half-life of the drug and its metabolite (M2), which accumulate is attributable to slow release from peripheral tissues. In multiple dosing over 2 weeks, there is about twofold increase in the area under the curve in 24 h ($AUC_{24}$) and the effective half-life is about 24 h [14].

In patients with mild-moderate renal and hepatic impairment, no dose adjustment is needed, but caution should be used with severe renal or hepatic impairment (no specific dose adjustment recommended). In population pharmacokinetic studies, Black subjects were found to have lower plasma concentrations than non-Black persons by ~34% of unknown mechanism but considered not clinically significant.

### 9.3.1 Drug-Drug Interactions of Bedaquiline

BDQ does not induce or inhibit CYP isoenzymes in vitro and does not affect the pharmacokinetics of most drugs. However, since BDQ is metabolized in the liver by CYP isoenzymes, potent inducers of CYP3A such as rifamycins (decrease the AUC by 40–59%) and others (efavirenz) should not be co-administered [14]. Co-administration of BDQ and moderate CYP3A4 inhibitors (e.g., ciprofloxacin, erythromycin, clarithromycin, fluconazole, ketoconazole, and ritonavir) for >14 days should be avoided [14].

### 9.3.2 Adverse Effects of Bedaquiline

It is difficult to tease out the adverse effects of BDQ, as it is always used in combination with other anti-TB agents for treatment of MDR-TB which may be responsible for these events. Adverse effects attributed to BDQ appear to be largely mild to moderate with gastrointestinal symptoms being the commonest, but QT prolongation on electrocardiogram (ECG) can occur; skin rashes, hyperuricemia, arthralgia, liver injury, and hyperlactatemia may be seen [16]. In long-term treatment with BDQ regimens ($n = 68$), adverse events were common but none required withdrawal of the drug, including seven (10.3%) with prolonged QT interval within 450–470 ms [17].

In a large retrospective study of 428 patients treated with BDQ, treatment was interrupted due to adverse events in 5.8%, 9.7% experienced QT prolongation of >500 ms, but the majority of adverse events included nausea, peripheral neuropathy, and vestibular toxicity, but 82.0% of patients also received linezolid which was likely responsible for the neuropathy [18].

### 9.3.3 Efficacy of Bedaquiline

The WHO treatment guidelines for drug-resistant TB have been revised in 2020 and include all oral shorter regimen containing BDQ [19]. The hierarchy of drugs for MDR-TB (at least resistant to INH and rifampin) and XDR-TB (resistant as well to the fluoroquinolones and 1 injectable drugs) are shown in Table 9.1. BDQ has been used in regimens for both MDR- and XDR-TB.

The efficacy of BDQ regimens have recently been reviewed and include 8 studies, 2 randomized and 6 cohort studies, with a total of 21,836 subjects [20]. This includes 1784 patients treated with BDQ and 20,061 not treated with BDQ, and 66.3% were human immunodeficiency virus (HIV) positive. BDQ was administered at 400 mg daily for 2 weeks and then 200 mg 3 times per week for 22 weeks, and

**Table 9.1** WHO category of drugs used for MDR-TB–2020 update

| Classification | Steps | Drugs |
|---|---|---|
| Group A | Include all three classes | Levofloxacin or moxifloxacin |
| | | Bedaquiline |
| | | Linezolid |
| Group B | Add one or both drugs | Clofazimine |
| | | Cycloserine or |
| | | Terizidone |
| Group C | Add to complete the regimen when A & B cannot be used | Ethambutol |
| | | Delamanid |
| | | Pyrazinamide |
| | | Imipenem or |
| | | Meropenem with |
| | | Clavulanic acid |
| | | Amikacin or |
| | | Ethionamide or |
| | | Prothionamide |
| | | p-Aminosalicylic acid |

Note: injectables being phased out
Adapted from Ref. [40]

duration of treatment was >6 months. BDQ was found to increase culture conversion (Relative Risk [RR] 1.272, $p < 0.0001$) and decrease the risk of all-cause mortality (RR: 0.529, p,0.001) but did not increase treatment success.

## 9.4 Delamanid

Delamanid (marketed as Deltyba) is a new anti-TB medication approved in several countries, including Japan and those of the European Union (EU), for treatment of MDR-TB in combination with other agents. In 2014, the WHO recommended its use and expand its use in children in 2016 for treatment of MDR-TB [21]. Delamanid has been made available to over 100 low- and middle-income countries through the Global Fund to Fight AIDS, TB, and malaria [22].

The drug is a first-in-its-class bicyclic nitroimidazole that inhibits the synthesis of methoxy-mycolic acid and keto-mycolic acid, cell wall components of mycobacteria [23]. Delamanid is a prodrug which is activated by the mycobacteria reductive metabolism to produce an active free radical, and the reactive intermediates of its metabolic pathway may affect the bacteria cellular respiration [24].

### 9.4.1 In Vitro Activity of Delamanid

Delamanid is equally active against both sensitive and MDR-TB strains with MICs between 0.006 and 0.012 μg/ml [24] and its early bactericidal effect in vivo appears to be similar to rifampin [25]. In a study from China, 220 strains of *M. tuberculosis* (110 MDR and 110 XDR) were tested against delamanid and pretomanid [26]. Delamanid was found to be more potent than pretomanid with $MIC_{90} \leq 0.016$ μg/ml, fourfold lower than pretomanid of 0.063 μg/ml. At a suggested breakpoint of 0.2 μg/ml [27], delamanid resistance was found in 7 (3.2%) of the isolates. In mice and human studies, delamanid showed dose-dependent killing [23].

DST is not standardized and not widely implemented. The WHO recommended critical concentration for delamanid DST was 0.016 μg/ml by Middlebrook 7H11 and 0.6 μg/ml by MGIT liquid culture [24]. MGIT (BD) is the proposed reference method for delamanid DST.

### 9.4.2 Resistance to Delamanid

Baseline resistance to delamanid has varied, depending on the country and the studies. In phase II and III studies, baseline resistance was very low (0.39% to 0.63%) and was acquired only in 1.17–1.95% [24]. In China, the rates of baseline resistance were 3.2–4.4% [26, 28], but was 9.67% in 420 strains from South Korea [29]. Data on acquired delamanid resistance in large treated groups besides phase II and III studies are lacking. In a prospective observational study of 156 MDR-TB cases, 31 received delamanid-based and 64 bedaquiline-based regimens [30]. Rates of acquired drug resistance were significantly higher in patients receiving delamanid versus bedaquiline (36% vs 10%, respectively, p,0.01). This was associated with lower sputum conversion rates and less favorable outcome with delamanid.

Mutations in five coenzyme F420 genes have been recognized in the laboratory and in clinical laboratory isolates (*fbiA*, *fbiB*, *fbiC*, *fgd*, and *Rv3547*) [24]. Loss of function mutation in *cofC* (*Rv2983*) gene was associated with delamanid and pretomanid resistance, and all pretomanid resistance with no mutations in the five previously reported genes [31]. High level resistance between delamanid and pretomanid may result from mutation in the *fbiA* gene, and in general cross-resistance between the two drugs was commonly found [24].

### 9.4.3 Pharmacokinetics of Delamanid

The relative bioavailability of different doses of delamanid varies from 58% to 76%, being higher with the lower dose, which may explain the non-proportional increase in the area under the concentration-time curve ($AUC_{0-24}$) with increased dosing, two-fold increase from 100 mg twice daily (BID) to 200 mg BID, resulted in a 1.5-fold increase in $AUC_{0-24}$ [32]. The recommended dose being 100 mg BID for 6 months and double the dose did not improve the outcome or sputum conversion [33]. Relative bioavailabilities in patients from Southeast Asia and Northeast Asia were 40% and 53% higher than patients from non-Asian regions, and absorption is significantly increased by taking with food, especially high-fat meal. Average peak plasma concentrations of delamanid were 0.2–0.6 μg/ml.

The elimination half-life was about 30–38 h, and albumin is responsible for metabolizing delamanid to its main metabolite and to a much less extent by cytochrome P450 enzymes. It is highly protein bound (99.5%) and hypoalbuminemia increased its clearance [32]. No dosage adjustment is recommended for mild to moderate renal impairment or liver dysfunction. Delamanid drug exposure was not affected by optimized background regimen for MDR-TB, lamivudine, tenofovir, CYP3A4 inhibitors and inducers, and antacids, but efavirenz increased its clearance by 35% [32].

### 9.4.4 *Efficacy and Safety of Delamanid*

In the phase II randomized, placebo-controlled, multinational clinical trial of 481 patients with MDR-TB of the lungs, 161 patients received delamanid 100 mg BID and 160 patients 200 mg BID, or placebo for 2 months in combination with a background drug regimen according to WHO guidelines [33]. Patients on the lower dose of delamanid at 2 months had sputum culture conversion of 45.4% versus 29.6% for the placebo group ($p = 0.04$), and the higher dose delamanid regimen was not more effective. Most adverse events were mild to moderate with similar rates among the groups, but QT prolongation without clinical consequences occurred in 9.9–13.1% in the delamanid groups and 3.8% in the placebo group.

A phase III randomized, double-blind, placebo-controlled trial was done in 7 countries in 511 patients with pulmonary MDR-TB [33]. Patients ($n = 341$) were assigned delamanid (100 mg BID for 2 months, then 200 mg OD for 4 months) or placebo ($n = 170$) with the same background regimen. The median time for sputum conversion rates between the 2 groups were similar (51 and 57 days) and adverse events between the groups were also similar (26.1% and 27.6%). Treatment-related deaths were similar between the groups (4.4% and 3.5%), but none were related to delamanid.

## 9.5 Pretomanid

Pretomanid is the second nitroimidazole prodrug developed for treatment of MDR-TB (after delamanid), and it belongs to the nitroimidazopyrans class [34]. It received FDA approval in combination with bedaquiline and linezolid for treatment of XDR-TB of the lungs, and treatment-intolerant or nonresponsive MDR-TB in 2019 (FDA news release). Pretomanid undergoes reductive "activation" within the mycobacteria, and it blocks the formation of keto-mycolic acids, a component of the cell wall, and the reactive nitro intermediates can also interrupt cellular respiration (similar to delamanid) [34]. Thus, under aerobic conditions, it inhibits cell wall synthesis through blockage of mycolic acid biosynthesis, but under anaerobic conditions it may kill non-replicating mycobacteria through nitric acid generation.

### 9.5.1 In Vitro Activity of Pretomanid

Pretomanid antimicrobial activity appears to be specific for mycobacteria (drug-susceptible, MDR- and XDR-TB) with MIC range from 0.05–0.48 μg/ml, but less active in vitro than delamanid [24]. The minimum bactericidal activity (MBC) is twice the MIC in aerobic conditions, while non-replicating mycobacteria in anaerobic conditions showed that the MBC was 7.5-fold the MIC [34]. The intracellular potency of pretomanid (in macrophages) is comparable to isoniazid but inferior to delamanid and rifampin. However, the early bactericidal activity in humans after 14 days of dosing was similar to isoniazid, rifampin, pyrazinamide, and ethambutol containing regimens [34].

Among clinically significant nontuberculous mycobacteria, pretomanid has significant in vitro activity only against *Mycobacterium bovis* and *Mycobacterium africanum* (MIC <0.031 to 0.125 μg/ml) [34].

### 9.5.2 Resistance to Pretomanid

In vitro, spontaneous mutation occurred at a frequency of $10^{-5}$ to $10^{-7}$, greater than rifampin but comparable to isoniazid, ethambutol, and pyrazinamide [35]. Resistance to pretomanid in animal models have also been described, but there is limited data in humans. Resistant mutations usually occur in the same genes responsible for its activity, *Ddn, Fgd*, and the proteins involved in $F_{420}$ biosynthesis (*fbiA*, *fibiB*, *fbiC*). Mutations of these five genes have been related to large increase in pretomanid MICs (>13- to >27-fold increase) [36]. Although cross-resistance between pretomanid and delamanid are often present, this is not necessarily so. Some mutations in the deazaflavin-dependent nitroreductase enzyme are associated with resistance to pretomanid but not to delamanid [37]. The critical concentration of pretomanid has been provisionally set at 1 μg/ml and over 99% of clinical isolates showed MIC values below this value (Pretomanid FGK, INN-pretomanid-europa.eu).

### 9.5.3 Pharmacology of Pretomanid

Pretomanid is highly lipophilic and is expected to diffuse across lipid membranes readily. Although in primates, pretomanid bioavailability appeared to be <50%; in healthy volunteers there was inter-individual variability with time of peak concentration ($T_{max}$) at 4–5 h [34]. Bioavailability in the fasting state was about half that in the fed state (best absorbed after high-fat meal); rate of absorption and bioavailability changed with the dose, reduced with increasing dose in the fasting state, but not in the fed state for doses <200 mg [38].

The drug is moderately high protein bound (86.5%) but widely distributed throughout the body, with volume of distribution of 92–180 L and penetrates the central nervous system (CNS) [34]. The median $C_{max}$ was 3.2 μg/ml and the median half-life was 18 h [38]. About 53% of the total dose appears in the urine (1% unchanged) and 38% is excreted in the feces.

Pretomanid undergoes extensive metabolic transformation by multiple metabolic pathways most abundant in the liver. The cytochrome P450 enzymes account for 20% of the biotransformation [34]. Co-administration with the potent CYP3A4 inducer rifampin reduces pretomanid exposure by >50%, and efavirenz reduces it by about 30%, but <20% with weak inducers such as lopinavir/ritonavir [34].

The recommended dose of pretomanid is 200 mg daily for 26 weeks. There is no data and guidelines yet on the effect of moderate to severe renal and hepatic impairment.

### 9.5.4 Clinical Efficacy and Safety of Pretomanid

Seven preclinical randomized controlled studies assessed the 2-week early bactericidal activity or 8-week bactericidal activity against rifampin-susceptible TB (6 studies) and rifampin-resistant TB (1 study) [39]. Pretomanid/moxifloxacin/pyrazinamide regimen was superior to standard therapy for time to culture conversion in rifampin-susceptible TB. In rifampin-resistant TB, the pretomanid regimen was not compared to other regimens but showed similar 8-week bactericidal activity as with the rifampin-susceptible TB.

The approval of pretomanid for XDR-TB was based on a single-group, open-label study from South Africa with a three-drug regimen of pretomanid (200 mg daily), BDQ (400 mg daily for 2 weeks, then 200 mg thrice weekly for 24 weeks), and linezolid (1200 mg daily) for 26 weeks (NIX TB trial) [40]. At the end of treatment, 90% of patients had a favorable outcome which was unprecedented for therapy of XDR-TB with the average rate of success being 14%. The unfavorable outcomes included seven deaths, two relapses, and one loss to follow-up. Side effects were common including peripheral neuropathy (81%) and myelosuppression (48%), largely attributed to linezolid.

The safety of pretomanid was difficult to assess in multidrug regimens, but in phase 1 studies in healthy volunteers and phase 2 studies, they were mild and included nausea, vomiting, rash, generalized pruritus, headaches, diarrhea, dizziness, and decreased hemoglobin (TB Alliance Data). Increased liver enzymes (7–13%) and serious hepatic adverse events (3%) were reported in short-term studies of bactericidal activity [39]. As noted in this review, 23/203 (11.3%) receiving pretomanid regimen had alanine transaminase >5 times upper limit of normal, compared to 4/68 (5.9%) patients on standard treatment, and 3 deaths on the experimental regimen were attributed to hepatotoxicity [39]. The role of pretomanid in liver toxicity, as both moxifloxacin and pyrazinamide can produce liver disturbance, is unclear but may be contributory. Prolonged QT interval has been found in some of the studies without clinical sequelae, but these changes could be secondary to the concomitant moxifloxacin or BDQ.

## 9.6 Conclusion and Discussion

The approval of these three new drugs, BDQ, delamanid, and pretomanid, represents a significant advance in the fight against drug-resistant TB. BDQ should now be included in all regimens for MDR-TB and XDR-TB, but trials are still ongoing to determine the best all oral short-term regimen for MDR-TB. Analysis of data from South Africa compared an all-oral BDQ-containing regimen ($n = 688$) to an injectable regimen ($n = 699$) for 9–12 months found higher success rates with the BDQ regimens (70% versus 57%) [41]. However, all these regimens contain the WHO-approved backbone of at least five other drugs. An ideal all-oral shorter regimen should be 3–4 drugs for 6 months.

The BDQ/pretomanid/linezolid regime used in the Nix-TB study [40] for XDR-TB was not sanctioned by the WHO for routine programmatic use worldwide in 2020, until more evidence are available for efficacy and safety [19]. This regimen may be used for MDR-TB patients with fluoroquinolone-resistant strains with no previous use of BDQ in research settings. The high prevalence of neuropathy and myelosuppression noted in the Nix-TB trial [40] was related to the 1200 mg daily dose of linezolid. A recent randomized trial showed that 600 mg linezolid daily for 26 weeks was as effective as 1200 mg daily for the same duration but with less myelosuppression (2% vs 22% with 1200 mg) and neuropathy (24% vs 38% with 1200 mg) and no optic neuropathy (0% vs 9% with 1200 mg) [42]. In May 2022, the WHO changed their recommendation for MDR-TB to include BDQ/pretomanid/moxifloxacin and 600 mg linezolid for 6 months (age > 14 years of age) [43].

There was sparse data on the efficacy and safety of a BDQ/delamanid-containing regimens for MDR-TB until recently. Data from observational study (EndTB study) and a randomized trial (DELIBERATE trial) had a total of 119 patients receiving this combination with other agents [44]. There was insufficient evidence to assess efficacy, but their concurrent use did not indicate additional safety issues above their individual use. Both BDQ and delamanid can cause prolonged QT interval which raised concerns about poor risk-benefit ratio. Recently, a multicenter, prospective observational cohort study of 472 patients with MDR-TB treated with concomitant BDQ/delamanid and other second-line drugs was reported [45]. Most patients (90.2%) had extensive disease and 74.2% had resistance to the fluoroquinolones; linezolid was used in 89.6% and clofazimine in 84.5%. The most common adverse events were peripheral neuropathy (28.4%), electrolyte disturbance (19.9%), acute kidney injury (8.5%), and myelosuppression (5.1%), and QT prolongation was only found in 1.5%. Treatment success occurred in 78.0%, 8.9% died, and treatment failure occurred in 7.2%. Thus, BDQ/delamanid concomitantly with other agents is safe and effective for MDR-TB. In a rare case, simultaneous resistance to BDQ and delamanid was reported from Japan with two nucleoside insertions (Rv0678 and fbiC) [46].

Future studies for MDR-TB should assess the combination of BDQ, pretomanid, or delamanid with pyrazinamide with or without a fluoroquinolone for 6 months. For XDR-TB, future studies should use tedizolid instead of linezolid with BDQ and pretomanid, as it appears to be safer and more active in vitro.

## References

1. World Health Organization. Global tuberculosis report 2019. Geneva: World health Organization; 2019. https://www.who.int/tb/publications/global=-report/en
2. Ramchandani SR, Gupta A, Bensen CA, et al. One month of rifapentine plus isoniazid to prevent HIV-related tuberculosis. N Engl J Med. 2019;380:1001–11.
3. Dorman SE, Nahid P, Kurbata EV, et al. Four—month rifapentine regimens with or without moxifloxcacin for tuberculosis. N Engl J Med. 2021;384:1705–18.
4. Khoshnood S, Goudarzi M, Taki E, et al. Bedaquiline: current status and future perspectives. J Global Antimicrob Resistance. 2021;25:48–59.
5. Matteelli A, Carvalho AC, Dooley KE, Kritski A. TMC207: the first compound of a new class of potent anti-tuberculous drugs. Future Microbiol. 2010;5:849–58.
6. Andries K, Verhasselt P, Guillemont J, et al. A diarylquinoline drug active on the ATP synthase of *Mycobacterial tuberculosis*. Science. 2005;307:223–7.
7. Chahine EB, Karaoui LR, Mansour H. Bedaquiline: a novel diarylquinoline for multidrug-resistant tuberculosis. Annn Pharmacother. 2014;48:107–15.
8. Andries K, Villellas C, Coeck N, et al. Acquired resistance of *Mycobacterium tuberculosis* to bedaquiline. PLoS One. 2014;9:e102135.
9. Ghajavand H, Kamakoli MK, Khanipour S, et al. High prevalence of bedaquiline resistance in treatment-naïve tuberculosis patients and verapamil effectiveness. Antimicrob Agents Chemother. 2019;63:e02530–18.
10. Nguyen TVA, Anthony RM, Banuls AL, Nguyen TVA, Vu DH, Alffenaar JWC. Bedaquiline resistance: its emergence, mechanism, and prevention. Clin Infect Dis. 2018;66:1625–30.
11. Bloomberg GV, Keller PM, Stucki D, et al. Acquired resistance to bedaquiline and delamanid in therapy for tuberculosis. N Engl J Med. 2015;373:1986–8.
12. Villellas C, Coeck N, Meehan CJ, et al. Unexpected high prevalence of resistance-associated *Rv0678* variants in MDR-TB patients without documented prior use of clofazimine or bedaquiline. J Antimicrob Chemothetr. 2016;72:684–90.
13. Liu Y, Gao M, Du J, et al. Reduced susceptibility of *Mycobacterium tuberculosis* to bedaquiline during antituberculous treatment and its correlation with clinical outcomes in China. Clin Infect Dis. 2021;73:e3391–7.

14. van Heeswijk RPG, Dannemann B, Hoetelmans RMW. Bedaquiline: a review of human pharmacokinetics and drug-drug interactions. J Antimicrob Chemother. 20–14; 69: 2310-8.
15. Upton CM, Steele C, Maartens G, Diacon AH, Wiesner L, Dooley KE. Pharmacokinetics of bedaquiline in cerebrospinal fluid [CSF] in patients with pulmonary tuberculosis [TB]. J Antimicrob Chemother. 2022;77:dka067. https://doi.org/10.1093/jaqc/dkac067.
16. Gaida R, Truter I, Peters CA. Adverse effects of bedaquiline in patients with extensively drug-resistant tuberculosis. S Afr J Infect Dis. 2020;35:23.
17. Olayanju O, Limberis J, Esmail A, et al. Long-term bedaquiline-related treatment outcomes in patients with extensively drug-resistant tuberculosis from South Africa. Eur Respir J. 2018;51:18005444.
18. Borisov SE, Dheda K, Enwerem M, et al. Effectiveness and safety of bedaquiline-containing regimens in the treatment of MDR- and XDR-TB: a multicenter study. Eur Respir J. 2017;49:1700387.
19. Mirzayev F, Viney K, Linh NN, et al. World Health Organization recommended treatment of drug-resistant tuberculosis, 2020 update. Eur Respir J. 2021;57:20003300.
20. Wang M-G, Wu S-Q, He J-Q. Efficacy of bedaquiline in the treatment of drug-resistant tuberculosis: a systemic review and meta-analysis. BMC Infect Dis. 2021;21:970.
21. World health Organization. WHO best-practice statement on the off-label use of bedaquiline and delamanid for the treatment of multidrug-resistant tuberculosis. Geneva: World Health Organization; 2017. p. 2017.
22. WHO. WHO consolidated guidelines on drug-resistant tuberculosis treatment. Available at: https://www.who.int/tb/publications/2019/consolidated-guiidelines-drug-resistant-TBtreatment/en/. Accessed 23 Nov 2021.
23. Matsumoto M, Hashizume H, Tomishinge T, et al. OPC-67683, a nitro-dihydro-imidazooxazole derivative with promising action against tuberculosis in vitro and in mice. PLoS Med. 2006;3:e466.
24. Nguyen TVA, Anthony RM, Cao TTH, et al. Delamanid resistance: update and clinical management. Clin Infect Dis. 2020;71:3252–9.
25. Diacon AH, Dawson R, Hanekom M, et al. Early bactericidal activity of delamanid [OPC-67683] in smear-positive pulmonary tuberculosis patients. Int J Tuberc Lung Dis. 2011;15:949–54.
26. Wen S, Jing W, Zhang T, et al. Comparison of in vitro activity of the nitroimidazoles delamanid and pretomanid against multi-drug resistant and extensive drug-resistant tuberculosis. Eur J Clin Microbial Infect Dis. 2019;38:1293–6.
27. Simson K, Kurepina N, Venter A, et al. MIC of delamanid [OPC-67683] against *Mycobacterium tuberculosis* clinical isolates and a proposed critical concentration. Antimicrob Agents Chemother. 2016;60:3316–22.
28. Pang Y, Zong Z, Huo F, et al. In vitro susceptibility of bedaquiline, delamanid, linezolid, clofazimine, moxifloxacin, and gatifloxacin against extensively drug-resistant tuberculosis. Antimicrob Agents Chemother. 2017;61:e00900–17.
29. Yang JS, Kim KJ, Choi H, Lee SH. Delamanid, bedaquiline, and linezolid minimum inhibitory concentration distributions and resistance related gene mutations in multidrug-resistant tuberculosis in Korea. Ann Lab Med. 2018;38:563–8.
30. Kempker RR, Mikiashvilli L, Zhao Y, et al. Clinical outcomes among patients with drug-resistant tuberculosis receiving bedaquiline- or delamanid-containing regimens. Clin Infect Dis. 2020;71:23336–44.
31. Rifat D, Li SY, Ioerger T, et al. Mutations in RV2983 as a novel determinant of resistance to nitroimidazole drugs in *Mycobacterium tuberculosis*. bioRxiv. 2018:457754. https://doi.org/10.1101/457754.
32. Wang X, Mallikaarjun S, Gibiansky E. Population pharmacokinetic analysis of delamanid in patients with pulmonary multidrug-resistant tuberculosis. Antimicrob Agents Chemother. 2021;65:e01202-20.

33. Gler MT, Skripconoka V, Sanchez-Garvito E, et al. Delamanid for multidrug-resistant tuberculosis. N Engl J Med. 2012;366:2151–60.
34. Stancil SL, Mirzayev F, Abdel-Rahman SM. Profiling pretomanid as a therapeutic option for TB infection: evidence to date. Dovepress. 2021;15:2815–30.
35. McGrath M, Gey van Pitius NC, van Helden PD, Warren RM, Warner DF. Mutation rate and the emergence of drug resistance in *Mycobacterium tuberculosis*. J Antimicrob Chemother. 2014;69:292–302.
36. Kadura S, King N, Nakhoul M, Zhu H, Theron G, Koser C, Farhal M. Systemic review of mutations associated with resistance to the new and repurposed *Mycobacterium tuberculosis* drugs bedaquiline, clofazimine, linezolid, delamanid and pretomanid. J Antimicrob Chemother. 2020;75:2031–43.
37. Lee BM, Almeida DV, Afriat-jurnou L, et al. The evolution of nitroimidazole antibiotic resistance in *Mycobacterium tuberculosis*. bioRxiv. 2019;63:631127.
38. Salinger DH, Subramoney V, Everitt D, Nedelman JR. Population pharmacokinetics of the antituberculous agent pretomanid. Antimicrob Agents Chemother. 2019;63:e00907–19.
39. Glis T, Lynen L, de Jung BC, Van Duen A, Decroo T. Pretomanid for tuberculosis: a systemic review. Clin Microbial Infect. 2021; https://doi.org/10.1016/j.c.mi.2021.08.007.
40. Conradie F, Diacon AH, Ngubane N, et al. Treatment of highly drug-resistant pulmonary tuberculosis. N Engl J Med. 2020;382:893–902.
41. Ndjeka N, Campbell JR, Meintjes G, et al. Treatment outcomes 24 months after initiating short, all-oral bedaquiline-containing or injectable-containing rifampicin-resistant tuberculosis treatment regimens in South Africa: a retrospective cohort study. Lancet Infect Dis. 2022. Published online May 2, 2022;22:1042. https://doi.org/10.1016/S1473-3099[21]00811-2.
42. Conradie F, Bagdasaryan TR, Borisov S, et al. Bedaquiline-pretomanid-linezolid regimens for drug-resistant tuberculosis. N Engl J Med. 2022;387:810–22.
43. World Health Organization. Rapid communication: key changes to the treatment to the treatment of drug-resistant tuberculosis. May 2, 2022. https://www.whoint/publications/i/item/WHO-UCN-TB-2022-2
44. Hewison C, Khan U, Bastard M, et al. Safety of treatment regimens containing bedaquiline and delamanid in the endTB cohort. Clin Infect Dis. 2022. Published online Jan 13;75:1006. https://doi.org/10.1093/cid/ciao19.
45. Huerga H, Khan U, Bastard M, et al. Safety and effectiveness outcomes from a 14-country cohort of patients with multi-drug resistant tuberculosis treated concomitantly with bedaquiline, delamanid and other second-line drugs. Clin Infect Dis. 2022:ciac176. https://doi.org/10.1093/cid/ciac176.
46. Yoshiyama T, Takaki A, Aono A, Mitarai S, Okumura M, Ohta K, Kato S. Multidrug resistant tuberculosis with simultaneous acquired drug resistance to bedaquiline and delamanid. Clin Infect Dis. 2021;73:2329–31.

# Part II
# New Systemic Antifungal Agents

# Chapter 10
# New Systemic Antifungal: Isavuconazole

## 10.1 Introduction

Invasive fungal infections are estimated to cause about 1.5 million deaths per year globally [1] and are projected to increase with increasing use of immunosuppressive drugs for cancer, organ, stem cell and bone marrow transplantations, and autoimmune diseases. Human immunodeficiency virus (HIV) is still an important cause of immunosuppression and the leading predisposition for cryptococcal meningitis. Despite current antifungal therapy, many invasive fungal infections result in mortality exceeding 50%. Moreover, with increased use of antifungal prophylaxis in many cancer centers there, has been increased prevalence of azole-resistant *Aspergillus* and *Candida* species.

Other reasons for the development of new antifungals include the shortcomings of the present agents. The current antifungals include four classes of drugs: the polyenes, azoles, echinocandins, and flucytosine. The polyene amphotericin-B is available only for parenteral use and has significant toxicities, especially renal, and the liposomal and lipid complex preparations are very expensive with restricted use. The azoles (fluconazole, itraconazole, voriconazole, and posaconazole) are commonly used but increased resistance of *Candida* species (fluconazole), erratic absorption (itraconazole and posaconazole), and serious drug-drug interactions (voriconazole and posaconazole) are shortcomings. Moreover, emergence of azole-resistant *Aspergillus* species have been increasing, mainly in cancer centers. Echinocandins (caspofungin, anidulafungin, micafungin) are only available for intravenous therapy and used for systemic invasive candidiasis, but *Candida auris* may pose a challenge with increased multidrug resistance. Flucytosine has limited use with amphotericin for cryptococcal meningitis.



I. W. Fong, *New Antimicrobials: For the Present and the Future*, Emerging Infectious Diseases of the 21st Century,
https://doi.org/10.1007/978-3-031-26078-0_10

## 10.2 Isavuconazole

Isavuconazole (isavuconazonium sulfate) is the latest azole to be approved by the United States (US) Food and Drug Administration (FDA) in 2015 (marketed as Cresemba) for treatment of invasive aspergillosis and mucormycosis. Isavuconazonium sulfate is the highly water-soluble prodrug that is hydrolyzed by plasma esterase into the active moiety (isavuconazole) [2]. The structure of isavuconazole (ISA) includes a side chain that orients the molecule to engage the triazole ring to the binding pocket of the fungal CYP51 protein, conferring broader antifungal activity than other azoles [3]. ISA inhibits the synthesis of ergosterol which leads to alterations in the structure and function of the fungal cell membrane leading to cell death.

## 10.3 In Vitro Activity

ISA has broad antifungal activity against yeasts, molds, and dimorphic fungi. It is active against most *Candida* species including *Candida krusei* and *Candida glabrata.* Comparison of the in vitro activity of ISA versus voriconazole against 1677 clinical *Candida* isolates (*C. albicans*, *C. dubliniensis*, *C. glabrata*, *C. krusei*, *C. parapsilosis*, *C. tropicalis*, and others) were comparable [4]. ISA was found to have lower minimum inhibitory concentrations (MIC) than fluconazole, but similar activity to amphotericin B, itraconazole, and voriconazole against 296 *Candida* blood isolates [5]. The $MIC_{50}$ was <0.5 μg/ml and the $MIC_{90}$ < 2 μg/ml. ISA activity against *C. auris* is variable and the echinocandins have better activity [6]. It is also very active against *Cryptococcus neoformans* and *Cryptococcus gattii* with MIC ranging from <0.008 to 0.5 μg/ml [7].

ISA is active against a wide range of *Aspergillus* species including species resistant to amphotericin B (*Aspergillus terreus*) and voriconazole/itraconazole-resistant species (*Aspergillus lentulus*) [8]. In a study of 702 strains of *Aspergillus* isolates, ISA $MIC_{90}$ was 1 μg/ml [9]. Susceptible breakpoint being ≤1 μg/ml and resistant >1 μg/ml [4]. Tables 10.1, 10.2, and 10.3 summarizes the in vitro activity of ISA.

Comparative in vitro activity of amphotericin B, voriconazole, posaconazole, and ISA has been reported to 72 clinical isolates of *Mucorales* [10]. All isolates were susceptible to amphotericin B, resistant to voriconazole, and more isolates appeared susceptible to ISA compared to posaconazole. However, ISA MICs were in general 1–3-fold higher than posaconazole, which may be compensated for by the higher blood levels with usual dosing. ISA demonstrated in vitro activity against most *Mucorales* isolates except for *Mucor circinelloides.*

ISA is also active against dimorphic fungi with MICs of 0.12 μg/ml to 2.0 μg/ml for *Coccidioides* species and *Histoplasma capsulatum* and for 6 isolates of *Blastomyces dermatitidis* (MIC 0.5–4 μ) [11].

**Table 10.1** In vitro activity of isavuconazole compared to fluconazole for *Candida* and *Cryptococcus* species

| Organisms | No. of isolates | Isavuconazole | | Fluconazole | |
|---|---|---|---|---|---|
| | | $MIC_{50}$ | $MIC_{90}$ | $MIC_{50}$ | $MIC_{90}$ |
| | | (μg/ml) | | (μg/ml) | |
| Candida species | | | | | |
| *C. albicans* | 744 | 0.008 | 0.016 | 0.5 | 1 |
| *C. glabrata* | 312 | 0.5 | 2 | 8 | 32 |
| *C. krusei* | 56 | 0.5 | 1 | 32 | 64 |
| *C. parapsilosis* | 285 | 0.06 | 0.125 | 1 | 2 |
| *C. tropicalis* | 155 | 0.06 | 0.125 | 1 | 1 |
| Other species | 102 | 0.016 | 0.25 | 1 | 16 |
| Cryptococcus species | | | | | |
| *C. neoformans* | 484 | 0.004 | 0.016 | 2 | 4 |
| *C. gattii* | 406 | 0.063 | 0.125 | 4 | 8 |

**Table 10.2** In vitro activity of isavuconazole compared to voriconazole against *Aspergillus* species

| Aspergillus species | No. of isolates | Isavuconazole | | Voriconazole | |
|---|---|---|---|---|---|
| | | $MIC_{50}$ | $MIC_{90}$ | $MIC_{50}$ | $MIC_{90}$ |
| | | (μg/ml) | | (μg/ml) | |
| *A. fumigatus* | 926 | 0.5 | 1 | 0.5 | 1 |
| *A. flavus* | 454 | 0.5 | 1 | 1 | 1 |
| *A. nidulans* | 106 | 0.125 | 1 | NA | NA |
| *A. niger* | 218 | 1 | 2 | 1 | 1 |
| *A. terreus* | 390 | 0.25 | 0.25 | 1 | 1 |
| *A. versicolor* | 75 | 0.25 | 0.5 | NA | NA |

**Table 10.3** In vitro activity of isavuconazole compared to posaconazole for *Mucorales*

| Mucorales | No. of isolates | Isavuconazole | | Posaconazole | |
|---|---|---|---|---|---|
| | | $MIC_{50}$ | $MIC_{90}$ | $MIC_{50}$ | $MIC_{90}$ |
| | | (μg/ml) | | (μg/ml) | |
| *Rhizomucor* | 29 | 2 | 16 | 0.25 | 0.5 |
| *Absidia* | 80 | 1 | 8 | 0.25 | 1 |
| *Rhizopus* | 139 | 1 | 4 | 0.5 | 1 |
| *Mucor* | 77 | 4 | 16 | 0.5 | 2 |
| *Cunninghamella* | 18 | 2 | 16 | NA | NA |

NB. Data for Tables 10.1, 10.2, and 10.3 adapted from Ref. [15]

## 10.4 Resistance to Isavuconazole

Among 2635 clinical isolates of *Candida* and *Aspergillus* isolates, up to 15% of *C. glabrata, C. tropicalis*, and *A. fumigatus* demonstrated increased resistance to ISA (non-wild-type strains) [4]. Emergence of resistant strains can occur with repeated exposure to drugs within the azole class, and fluconazole resistance in *Candida* species can confer resistance to ISA [12]. Three main mechanisms of resistance to the azoles are: (i) mutations in the gene coding the target enzyme (ERG11) leading to decrease binding of azoles; (ii) mutations in the ERG3 gene resulting in the inability to disrupt the cell membrane; and (iii) overexpression of the efflux pumps through the ATP binding cassette (ABC) transporter [13]. Multiple mutations can exist in a single a *Candida* strain and lead to cross-resistance between the triazoles. The ABC transporters, CDr1 and CgCDR1, in *C. albicans* had the greatest effect on ISA MICs [13] but unlike fluconazole and voriconazole, were less affected by the major facilitator superfamily transporters MDR1 or FLU1.

Azole resistance in *Aspergillus* species is associated with alterations in the Cyp51A gene which result in changes in the enzyme targeted by the azoles; other mechanisms include efflux pumps and mutations in the promoter region of Cyp51 A [8, 12]. *Aspergillus* species with Cyp51 A mutations have cross-resistance to ISA and voriconazole and other azoles [14].

## 10.5 Pharmacology

ISA is available as the prodrug (isavuconazonium sulfate) in both oral and intravenous formulations. It is highly water soluble, and the intravenous formulation does not require cyclodextrin for solubilization, unlike voriconazole and posaconazole [11]. The cyclodextrin vehicle can be nephrotoxic. The prodrug is rapidly converted to the active component (ISA) and an inactive component after intravenous infusion.

The oral bioavailability of ISA is 98%, and it can be taken with or without food with maximum plasma concentration ($C_{max}$) in 2–3 h with 2.5 ± 1.0 μg/ml at steady state [11]. It is widely distributed in the body with a large volume of distribution (450 L), high protein binding (>99%), and a long terminal half-life of 100–130 h. Less than 1% of ISA is eliminated in the urine, and in animal studies it is primarily excreted in the feces.

Hepatic metabolism by the CYP enzyme family, CYP3A4 and CYP3A5 isoenzymes, are the primary means of elimination [11]. Dosage adjustment is not recommended for renal impairment or for mild to moderate hepatic impairment, and there is no data for severe liver disease. The total drug exposure as reflected by the area under the concentration curve (AUC) is increased by 85% for mild liver impairment and 159% with moderate impairment [15]. As with voriconazole, Chinese subjects compared to White subjects have lower clearance of ISA by about 50%, but no dosage adjustment is necessary [11].

ISA is found in breast milk of lactating animals and labeled as a class C drug and should not be given to pregnant women.

The recommended dosing of intravenous and oral formulations of ISA are similar, loading dose over the first 2 days as 200 mg every 8 h, followed by 200 mg daily [11]. This dosing regimen produced ISA serum concentration above the $MIC_{90}$ for

*Aspergillus* and the $MIC_{50}$ for *Mucorales* by day 1, and trough concentrations of 2–3 μg/ml before starting the 200 mg daily dose [15].

## 10.6 Drug-Drug Interactions

Significant drug-drug interactions occur with inhibitors and inducers of the CYP hepatic enzymes. Inhibitors of CYP3A4 enzyme result in increased levels of ISA and should be used with caution, while potent inducers (rifampin, carbamazepine, etc.) greatly reduce ISA blood levels and should be avoided; see Table 10.4 [11].

ISA moderately inhibits (CYP3A4) the metabolism of immune-modulators used in transplants (sirolimus, tacrolimus, and cyclosporine), leading to increased levels which should be monitored. However, compared to voriconazole and posaconazole, ISA has less drug-drug interactions [11].

## 10.7 Clinical Efficacy

### *10.7.1 Aspergillosis*

Despite advances in antifungal therapy, invasive aspergillosis still carries high mortality and morbidity, and in hematopoietic stem cell transplants (HSCT) the overall 1-year survival after treatment is only 25.4% [16]. In the SECURE trial, patients with invasive mold infections were randomized to receive voriconazole ($n = 264$) or

**Table 10.4** Isavuconazole drug-drug interactions

| Type of interaction | Drug | Recommendation |
|---|---|---|
| Increases ISA level | Lopinavir/ritonavir | Use with caution |
| Decreases ISA level | Rifampin | Contraindicated |
| | Carbamazepine | Avoid |
| | Long-acting barbiturates | Avoid |
| | St. John's wort | Avoid |
| Levels increased by ISA | Sirolimus/tacrolimus | Caution, monitor levels |
| | Cyclosporin | Caution, monitor levels |
| | Mycophenolate | Caution. monitor levels |
| | Digoxin | Caution. monitor levels |
| | Colchicine | Caution, may adjust dose |
| | Dabigatran, midazolam | Caution, may adjust dose |
| | Atorvastatin | None |
| Levels decreased by ISA | Bupropion | Caution, may increase dose |
| | Lopinavir/ritonavir | Caution |

NB. Data adapted from Ref. [11]

ISA ($n$ = 263), but of the 516 total patients, only 272 patients had proven (65 patients) or probable (207 patients) invasive mold infections [17]. The primary endpoint was all-cause mortality at day 42. This was a double-blind, noninferiority, phase 3 trial. The 42-day mortality for the ISA arm was 18.6% and was not significantly different from the voriconazole arm of 20.2% (noninferior). The overall success (complete and partial response) at the end of treatment was also similar for the two drugs, 35% for ISA and 34% for voriconazole [17]. ISA was associated with less adverse events, and treatment discontinuation due to these events was lower in the ISA arm.

In a single-center retrospective matched cohort of 100 patients treated for invasive fungal infections, the composite safety outcome was compared for ISA ($n$ = 33), voriconazole ($n$ = 34), and posaconazole ($n$ = 33) [18]. The composite safety outcome consisted of drug-related QTc prolongation, elevated liver enzymes (5 times upper limit of normal), or any documented adverse drug event. ISA produced significant lower adverse out comes (24.2%) compared to voriconazole [55.95], and posaconazole (39.4%, $p$ = 0.029), but the drug costs were comparable for the different agents. There is accumulating data on the safety and efficacy of ISA in immunosuppressed children with invasive aspergillosis, which showed good tolerance and response rate of 50% when used as first-line treatment [19].

Long-term antifungal therapy for 6–12 months has been used for a rare disorder in mostly immunocompetent patients, chronic pulmonary aspergillosis which produces slow destruction of the lungs with a high morbidity and mortality [20]. The drug of choice is itraconazole with voriconazole as an alternative. In a retrospective study of chronic pulmonary aspergillosis, comparison of rates of adverse events were assessed in patients treated with voriconazole ($n$ = 21) and ISA ($n$ = 20) for 6–12 months [21]. Adverse events occurred in 18 of 21 (86%) on voriconazole and 12 of 20 (60%) patients on ISA ($p$ = 0.02), but the rates of discontinuation were similar. Five (25%) patients in the ISA group were intolerant of other triazoles tolerated ISA.

### 10.7.2 Mucormycosis

Mucormycosis is a rare disease with fatality rates of up to 80–90% in immunocompromised patients, and treatment includes antifungal agents and surgical debridement. Antifungal treatment consists of amphotericin-B (including liposomal and lipid formulations) and posaconazole (usually for maintenance therapy). In developed countries, invasive mucormycosis is mostly commonly found in immunosuppressed patients, but diabetes is the main underlying disease globally. The prevalence of mucormycosis in India is about 80 times greater than in developed countries [22], and it is now recognized in that country as a complication of COVID-19 infection. Randomized trials are not feasible due to the rarity of the disease.

ISA was approved by the FDA for invasive mucormycosis based on the VITAL study. The VITAL study was an open-label non-comparative study of ISA in patients with invasive aspergillosis and renal impairment or with invasive fungal infections with rare fungi [23]. Thirty-seven patients had proven (86%) or probable (14%) invasive mucormycosis. ISA was the primary treatment in 21 (56.7%), 11 had refractory disease, and 5 were intolerant to other therapy. The main underlying condition was hematologic malignancy (59%) and 27% were neutropenic and pulmonary infection was present in 59%. The mortality at day 42 was 38% and the overall response rate was 31.4%, which were similar to those reported with amphotericin-B.

Patients ($n$ = 21) receiving primary treatment with ISA from the VITAL study were compared to 33 patients treated with amphotericin from the FungiScope Registry in a matched case-control analysis [23]. At day 42, the all-cause mortality was similar between the two treatments and survival up to day 84 was not significantly different. Six pediatric cases of mucormycosis treated successfully with ISA have been described in the literature: 4 of 6 as a rescue therapy and as part of combination antifungal therapy and most cases required surgical debridement [19].

### *10.7.3 Candidiasis*

Invasive candidiasis has been increasing in hospitals and *Candida* species are among the top three causes of hospital-associated infections with high mortality of 30–40% despite antifungal therapy. Guidelines have recommended echinocandins as first-line initial therapy, but no oral formulations are available for this group of agents and their superiority over azoles (i.e., fluconazole) is not well established.

The ACTIVE trial was a phase 3 double-blind clinical trial for treatment of candidemia or invasive candidiasis comparing intravenous (IV) and then oral ISA ($n$ = 199) to IV caspofungin and then oral voriconazole ($n$ = 201) for a minimum of 14 days after the last positive blood culture for up to 56 days [24]. Both groups received IV therapy for 10 days. Central catheter was removed for all patients with candidemia. The most common species in both arms were *C. albicans*, *C. tropicalis*, *C. parapsilosis*, and *C. glabrata*, but overall non-albicans species were the most common.

The primary endpoint with overall response at the end of IV therapy was 60.3% for ISA and 71.1% for caspofungin, with failure of ISA to demonstrate noninferiority to caspofungin. However, the overall response rates 2 weeks after end of treatment and survival on days 14 and 56 were similar. The clearance rate of candidemia was similar, and breakthrough or recurrent infections were slightly higher in the caspofungin group. In patients transitioned from IV to oral therapy, ISA was successful in 82.6% and caspofungin/voriconazole in 77.5%.

### 10.7.4 Other Fungal Infections

There is limited data on the use of ISA for treatment for rare mold infections and dimorphic fungal infections. Only 3 of 9 fusariosis (*Fusarium* species infection) and 1 of 3 scedosporiosis (due to *Scedosporium* species) had a complete or partial response to ISA [11]. Thus, ISA is not a promising agent for treatment of these rare mold infections.

ISA has been used in small numbers of patients with cryptococcosis or dimorphic fungal infections. In a report of 38 patients, 9 had cryptococcosis, 10 had paracoccidioidomycosis, 9 had coccidioidomycosis, 7 had histoplasmosis, and 3 had blastomycosis, treated for a median of 180 days [25]. Treatment response occurred in 24 (63%) patients and although adverse events occurred in 87% (many not likely drug related), ISA was not discontinued because of side effects.

Coccidioidal meningitis is a rare and difficult disease to treat, as patients require life-long antifungal therapy, and the cumulative drug toxicity and treatment failure often require salvage therapy. A recent report outlines the use of ISA in 9 patients with coccidioidal meningitis treated with a mean of 504 days [26]. ISA was used because of treatment failure in one patient and the others due to serious side effects of voriconazole, after primary treatment with fluconazole. Therapy was successful in three patients and six patients were stable with no relapse.

## 10.8 Safety

ISA has been well tolerated in various trials with over 1700 patients receiving the agent in phase I, II, and III trials [15]. In the SECURE trial, fewer drug-related adverse events were reported with ISA (42.4%) compared to voriconazole (59.8%). Most side effects are mild and gastrointestinal in nature, nausea, vomiting and diarrhea, and others include headache, rash, elevated liver enzymes, and shortening of the QTc interval (of unknown significance); less common side effects include peripheral edema, hypokalemia, and infusion reactions (chills, dyspnea, and hypotension) [11]. Unlike other triazoles, ISA does not cause QT interval prolongation which may be associated with ventricular arrhythmias.

## 10.9 Conclusion

ISA is a new triazole that is available in oral and IV formulations with excellent bioavailability and broad spectrum of activity that is approved for treatment of invasive aspergillosis and mucormycosis. It is not more effective than current triazoles, but it has less drug-drug interactions than voriconazole and posaconazole, is more reliably absorbed, can be administered once daily, and appears safer than the other triazoles. Thus, it could be used as primary or secondary treatment of invasive aspergillosis and mucormycosis. Its safety profile is especially advantageous for longer-term treatment as in chronic pulmonary aspergillosis and coccidioidal meningitis. Moreover, cost-effective analyses have shown that ISA is more cost-effective for the primary treatment of invasive aspergillosis than voriconazole [27, 28].

## References


1. Brown GD, Denning DW, Gow NA, Levitz SM, Netea MG, White TC. Hidden killers: human fungal infections. Sci Transl Med. 2012;4:165rv113.
2. Van Daele R, Spriet I, Wauters J, Maertens J, Mercier T, Van Hecke S, Bruggemann R. Antifungal drugs: what brings the future? Med Mycol. 2019;57(Suppl. 3):S328–43.
3. Jenks JD, Salzer HJ, Prattes J, Krause R, Buchheidt D, Hoenig M. Spotlight on isavuconazole in the treatment of invasive aspergillosis and mucormycosis: design, development, and place in therapy. Drugs Dev Ther. 2018;12:1033–44.
4. Astvad KMT, Hare RK, Arendrup MC. Evaluation of the *in vitro* activity of isavuconazole and comparator voriconazole against 2635 contemporary clinical *Candida* and *Aspergillus* isolates. Clin Microbiol Infect. 2017;23:882–7.
5. Seifert H, Aurbach U, Stefanik D, Cornerly O. In vitro activities of isavuconazole and other antifungal agents against *Candida* bloodstream isolates. Antimicrob Agents Chemother. 2007;51:1818–21.
6. Osei SJ. *Candida auris*: a systematic review and meta-analysis of current updates on an emerging multi-resistant pathogen. Microbiology. 2018;7:e00578.
7. Espaniel-Ingroff A, Chowdhary A, Gonzalez GM, et al. Multicenter study of isavuconazole MIC distribution and epidemiological cutoff values for *Cryptococcus neoformans-Cryptococcus gattii* species complex using the CLSI M27-A3 broth microdilution method. Antimicrob Agents Chemother. 2015;59:666–8.
8. Ellsworth M, Ostrosky-Zeichner L. Isavuconazole: mechanism of action, clinical efficacy, and resistance. J Fungi [Basel]. 2020;6:324.
9. Guinea J, Pelaez T, Recio S, Torres-Narbona M, Bouza E. In vitro antifungal activities of isavuconazole [BAL4815], voriconazole, and fluconazole against 1007 isolates of zygomycete, *Candida, Aspergillus, Fusarium, and Scedosporium* species. Antimicrob Agents Chemopther. 2008;52:1396–400.
10. Arendrup MC, Jensen RH, Meletiadis J. In vitro activity of isavuconazole and comparators against clinical isolates of the Mucorales order. Antimicrob Agents Chemother. 2015;59:7735–42.
11. Micel MH, Kauffman. Isavuconazole: a new broad-spectrum triazole antifungal agent. Clin Infect Dis. 2015;61:1558–65.
12. Sanglard D, Coste AT. Activity of isavuconazole and other azoles against *Candida* clinical isolates and yeast model systems with known azole resistance mechanisms. Antimicrob Agents Chemother. 2015;60:229–38.
13. Pfaller MA. Antifungal drug resistance: mechanisms, epidemiology, and consequences for treatment. Am J Med. 2012;125(Suppl. 1):S3–S13.
14. Jorgensen KM, Astvad KMT, Hare RK, Arendrup MC. EUCAST susceptibility testing of isavuconazole: MIC data for contemporary clinical mold and yeast isolates. Antimicrob Agents Chemother. 2019;63:e00073–19.
15. Rybak JM, Marx KR, Nishimoto AT, Rogers PD. Isavuconazole: pharmacology, pharmacodynamics, and current clinical experience with a new triazole antifungal agent. Pharmacotherapy. 2015;35:1037–51.

16. Kontoyiannis DP, Marr KA, Park BJ, et al. Prospective surveillance for invasive fungal infections in hematopoietic stem cell transplant recipients, 2001–2006: overview of the transplant-associated infection surveillance network [TRANSNET] database. Clin Infect Dis. 2010;50:1091–100.
17. Maertens JA, Raad II, Marr KA, et al. Isavuconazole versus voriconazole for primary treatment of invasive mold disease caused by *aspergillus* and other filamentous fungi [SECURE]: a phase 3 randomized-controlled, non-inferiority trial. Lancet. 2016;387:760–9.
18. Van Matre ET, Evans SL, Mueller SW, Maclaren R, Fish DN, Kiser TH. Comparative evaluation of isavuconazole, voriconazole, and posaconazole for the management of invasive fungal infections in an academic medical center. Ann Clin Microbiol Antimicrob. 2019;18:13. https://doi.org/10.1186/s12941-019-0311-3.
19. Decembrino N, Percuccio K, Zecca M, et al. A case series and literature review of isavuconazole use in pediatric patients with hemato-oncologic diseases and hematopoietic stem cell transplantation. Antimicrob Agents Chemother. 2020;64:e01783–19.
20. Barac A, Kosmidis C, Alastruey-Izquiierdo A, Salzer HJF, Anet CP. Chronic pulmonary aspergillosis update: a year in review. Med Mycol. 2019;57:S104–9.
21. Bongomin F, Maguire N, Moore CB, Felton T, Rautemaa-Richardson R. Isavuconazole and voriconazole for the treatment of chronic pulmonary aspergillosis: a retrospective comparison of rates of adverse events. Mycoses. 2019;62:217–22.
22. Skiada A, Pavleas I, Drogari-Apiranthitou M. Epidemiology and diagnosis of mucormycosis: an update. J Fungi [Basel]. 2020;6:265. https://doi.org/10.3390/jof6040265.
23. Marty FM, Ostrosky-Zeichner L, Comely OA, et al. VITAL and FungiScope Mucormycosis investigators. Isavuconazole treatment for mucormycosis: a single-arm open label trial and case-control analysis. Lancet Infect Dis. 2016;16:828–37.
24. Kullberg BJ, Viscoli C, Pappas PG, et al. Isavuconazole versus caspofungin in the treatment of candidemia and other invasive *Candida* infections: the ACTIVE trial. Clin Infect Dis. 2019;68:1981–9.
25. Thompson GR 3rd, Rendon A, Dos Santos RR, et al. Isavuconazole treatment of cryptococcosis and dimorphic mycoses. Clin Infect Dis. 2016;63:356–62.
26. Heidari A, Quinlan M, Benjamin DJ, et al. Isavuconazole in the treatment of coccidioidal meningitis. Antimicrob Agents Chemother. 2019;63:e02232–18.
27. Harrington R, Lee E, Yang H, Wei J, Messali A, Azie N, Wu EQ, Spalding J. Cost-effective analysis of isavuconazole vs. voriconazole for invasive aspergillosis. Adv Ther. 2017;34:207–20.
28. Floros L, Kuessner D, Posthumus J, Bagshaw E, Sjolin J. Cost-effective analysis of isavuconazole versus voriconazole for the treatment of patients with possible invasive aspergillosis. BMC Infect Dis. 2019;134:1. https://doi.org/10.1186/s12879-019-3683-2.

# Part III
# Anti-Parasitic Agents

# Chapter 11
# Newly Approved Anti-parasitic Drugs for Malaria, Fascioliasis, Onchocerciasis, Chagas Disease, and African Trypanosomiasis

## 11.1 Introduction

Parasitic diseases afflict hundreds of millions of people globally each year but the actual burden on global health has been difficult to estimate. The greatest charge from parasitic diseases is borne by populations in developing countries located in tropical and subtropical regions of the world. Despite their enormous toll on human health, development of new therapeutic drugs for these diseases has long been neglected. Between 1975 and 1999, only 13 of 1300 (0.01%) new therapeutic drugs introduced were slated for parasitic diseases [1]. However, since 2000 there has been 20 new anti-parasitic drugs introduced or in development, which is a great stride forward. This chapter reviews two old and three new anti-parasitic drugs approved by the Federal Food and Drug Administration (FDA) in the United States (US) since 2015: (i) benznidazole (old drug) for Chagas disease; (ii) fexinidazole for African trypanosomiasis; (iii) moxidectin for onchocerciasis; (iv) tafenoquine for malaria; and (v) triclabendazole (old drug) for fascioliasis.

## 11.2 Benznidazole

### *11.2.1 Chagas Disease*

Chagas disease, also known as American trypanosomiasis, caused by the protozoa, *Trypanosoma cruzi,* is a vector-borne zoonosis transmitted by the Triatominae insect ("kissing bug"), but occasionally transmitted by blood transfusion, orally (food-borne), vertically (mother-to-child), and organ transplantation. Worldwide, mostly in Latin America, it is estimated that six to seven million people are infected



I. W. Fong, *New Antimicrobials: For the Present and the Future*, Emerging Infectious Diseases of the 21st Century,
https://doi.org/10.1007/978-3-031-26078-0_11

with the parasite [2]. Chagas disease used to be limited to the rural areas of Latin America, but with increased migration in the last decade most infected people live in urban settings and the disease is now present in North America and Europe. In 2016 it was estimated that there were 238,000 cases of Chagas disease in the US [3], and studies in Europe showed an overall 4.2% prevalence, with the highest infection rates in people from Bolivia (18.1%) [4].

Infection with *T. cruzi* from the vector typically occurs in early childhood and the acute phase is usually asymptomatic but fever and malaise occur in about 5%, followed by a prolonged asymptomatic indeterminate phase (>50% remain in this phase for life with no sequelae); after a decade or more 35% to 50% enter a chronic phase with chronic heart disease (up to 30%) and intestinal disease (10% with megacolon and megaesophagus) [2, 5]. Congenital infection and reactivation in immunosuppressed patients, including those with AIDS, can occasionally lead to myocarditis and meningoencephalitis or focal brain lesions, and AIDS patients can also present with subcutaneous nodules [6]. The infection is estimated to cause 10,000 deaths every year [5].

### 11.2.2 *Treatment of Chagas Disease*

No new therapy has been introduced for Chagas disease since the 1970s, but benznidazole which has been used in Latin America for decades was only approved by the FDA in 2017. This is due to the increasing number of migrants from Central and South America now recognized to have Chagas disease in the US. The two available drugs for Chagas disease belong to the nitroimidazole class: nifurtimox was released in 1967, followed by benznidazole in 1972 [7]. Both drugs were very effective in the acute phase of Chagas disease and were generally not used in the chronic phase, as the persistence of the protozoa was not believed responsible for chronic disease but considered related to the immune response of the host. Benznidazole (BZN) is better tolerated than nifurtimox and is considered the first drug of choice by many experts [7].

BZN is a prodrug that is enzymatically activated by the trypanosomal type I nitroreductases to produce reactive metabolites which are toxic to cells and DNA, causing rapid trypanosomal effect of intracellular and extracellular parasite [8]. It is active against both the trypomastigote and amastigote forms [6]. The tablet is well absorbed (92% bioavailability) and not affected by food but the peak serum concentration occurs at 2 h in the fasting state and 3.2 h after meals, with maximum serum concentration (Cmax) of 2.4 μg/ml and half-life of 12 h (Exeltis drug manual). The volume of distribution was higher in men than women (125.9 versus 88.6 liters), which may account for lower Cmax (1.6 vs 2.9 μg/ml) [9]. BZN is excreted primarily by the kidneys and 22% by the fecal route [6]. The dose is 5–8 mg/kg/day in two divided doses for 60 days, and it is available as 12.5 mg and 100 mg tablets. It should not be given to pregnant women (in animals causes fetal toxicity) and should not be used in renal or hepatic failure [2].

### 11.2.3 Efficacy of Benznidazole

BZN is most effective in acute or early congenital infection to shorten clinical course and clear parasitemia, with parasitological cure of 60–85% in the acute phase and 90% of congenital infection in the first year [6]. For children <30 kg body weight, the 12.5 mg tablet can be made into a slurry with water.

The use of BZN in the chronic phase of Chagas disease has been controversial, but currently there is consensus that the persistence of the parasite is responsible for inducing persistent inflammation that leads to chronic disease progression. Hence, eradication of the trypanosome may be necessary to prevent progression from the intermediate phase to Chagas heart disease (cardiomyopathy). This is supported by the presence of the trypanosome antigen and the severity of myocardial inflammation in Chagas disease [9]. Previous randomized trials in children aged 6–12 years with asymptomatic chronic T. *cruzi i*nfection demonstrated that treatment resulted in seroconversion from positive to negative serology in 60% [6]. Thus, early diagnosis and BZN treatment was recommended for all children.

But does treatment of chronic Chagas disease produce any clinical benefit? In a multicenter, randomized study, 2854 patients with Chagas cardiomyopathy received BZN ($n = 1431$) or placebo ($n = 1432$) for 80 days [10]. Of 1896 patients with blood tested for *T. cruzi* by PCR, 60.5% were positive and rates of conversion to negative PCR were greater in the treated group at the end of treatment, 2 and 5 years post-treatment ($p < 0.001$), but the gap narrowed at 5 years to 46.7% versus 33.1% for BZN and placebo, respectively. However, there was no difference in the rate of clinical deterioration between the two groups.

What is the benefit of treatment in the indeterminate phase of Chagas disease without cardiac disease? There is no randomized study to prove the benefit of BZN in the indeterminate stage of Chagas disease, but several observation studies suggest that treatment may slow the progression to Chagas heart disease. In a retrospective observational cohort of 228 patients, 114 patients treated with BZN for 30–60 days were compared to matched non-treated patients with a median follow-up of 15.1 years [11]. The rate of progression was less with treatment, 21.1% of untreated patients progressed to cardiac form of Chagas compared to 7.9% of treated patients, $p = 0.04$.

The indications for treatment with BZN include all acute and congenital infections, reactivated infections (immunosuppressed), for children 18 years or younger with chronic infection, and women of child-bearing age (before or after pregnancy) to prevent congenital infection [2, 6]. Adults under 60 years of age in the chronic stage without advanced cardiac disease should also be offered treatment.

### 11.2.4 Adverse Effects of Benznidazole

BZN is genotoxic and can cause risk to a fetus and is contraindicated in pregnancy and during lactation. The most common side effects are gastrointestinal (GI) in nature, abdominal pain (up to 25%), nausea and vomiting (5%), decreased appetite, rashes (11–16%), occasionally dizziness, peripheral neuropathy, tremors, leucopenia, or neutropenia (see drug manual). However, most patients treated with nifurtimox experience side effects, 68.7% GI adverse events and 60.5% neurologic events (MMWR, March 11, 2022).

## 11.3 Fexinidazole

### *11.3.1 African Trypanosomiasis*

There are two forms of human African trypanosomiasis, depending on the subspecies of the parasite: (i) *Trypanosoma brucei gambiense* which causes 95% of cases of sleeping sickness and is found in 24 countries of west and central Africa. Infected patients can remain subclinical for months to years, and central nervous system (CNS) signs represent advanced disease. (ii) *Trypanosoma brucei rhodesiense* accounts for 5% of reported cases and is found in 13 countries in eastern and southern Africa [12]. Acute infection results in signs and symptoms in weeks to a few months with rapid involvement of the CNS.

With sustained control measures for many years, the number of cases fell below 10,000 for the first time in 2009, and only 992 new cases were recorded in 2019 [12]. Thus, eradication of the disease is in sight. Infection is transmitted by the bite of the tsetse fly with trypanosome from infected humans or animals. Infection that can be transmitted rarely by other means, vertically from mother to fetus, from contact with contaminated needles, and transmission by sexual contact has been reported [12].

Clinically the disease is divided into two stages: the first stage or hemolymphatic stage, the parasite multiply in the subcutaneous tissues, blood, and lymph, and patients present with fever, headaches, lymphadenopathy, joint pains, and pruritus; the second stage the trypanosome enters the CNS to cause the meningo-encephalic stage. The second stage is characterized by disturbance of sleep cycle, confusion, changes in behavior, sensory disturbance, and poor coordination.

### *11.3.2 Treatment of African Trypanosomiasis*

There are six drugs used in the treatment of African trypanosomiasis, and this depends on the stage of disease. Drugs used in the first stage: (i) pentamidine in the first stage of *T. b. gambiense* infection; (ii) and suramin used for the first stage of *T. b. rhodesiense* infection.

Drugs used in the second stage are the following: (iii) melarsoprol (arsenic compound) used as first-line treatment of *rhodesiense* form and rarely used in the *gambiense* form; (iv) eflornithine is only effective against *T. b. gambiense*; *(v)* nifurtimox is used in combination with eflornithine for the gambiense form. Then drugs used in both stages—(vi) fexinidazole.

Fexinidazole (FNZ) is indicated as first-line treatment of the first stage of *gambiense* form and non-severe second stage [12], and it was approved by the US FDA in 2021. FNZ is a 2-substituted 5-nitroimidazole discovered in the late 1970s but rediscovered >30 years later by the Drugs for Neglected Diseases initiative. FNZ and its two active metabolites (a sulfoxide and a sulfone) are active against the 2 human African trypanosoma species in vitro with the inhibitory concentration of 50% ($IC_{50}$) of clinical isolates ranging between 0.16 and 0.93 μg/ml [13]. The mechanism of action is unclear but it is believed that reduction of FNZ and its two active metabolites by trypanosome nitroreductase produce reactive intermediates that damage the protozoa DNA and proteins [14].

### 11.3.3 Pharmacology of Fexinidazole

FNZ is rapidly absorbed after oral dosing and rapidly metabolized by the cytochrome P450 enzymes to its active metabolites with maximum concentration at 2–5 h, and food increases the bioavailability greatly (Winthrop drug manual). The half-life of FNZ is about 11–14 h and the active metabolites 15 to 23 h, the protein binding is 95.4%, and the cerebrospinal fluid (CSF) concentration of the active metabolites in adults is about 31% to 52% of plasma concentration. Only <3.15% of the dose is excreted in the urine and most is excreted by the biliary route. The usual adult dose is 1800 mg (3 tablets) once daily with food for 4 days, then 1200 mg daily for 6 days; in children ≥20 and < 35 kg, the dose is 1200 mg daily for 4 days and then 600 mg daily for 6 days. No dose adjustment is needed for renal impairment, and no data is available for hepatic impairment, but it is contraindicated in patients with signs of cirrhosis or jaundice.

### 11.3.4 Clinical Efficacy of Fexinidazole

FNZ is the first exclusively oral treatment for human African trypanosomiasis (HAT). In a randomized, phase 2/3, noninferiority, open-label trial, 394 patients with late-stage HAT were randomized to oral FNZ ($n = 264$) or nifurtimox/eflornithine (NECT) combination ($n = 130$) [15]. FNZ was given once daily, 1800 mg for 4 days, and then 1200 mg for 6 days. Oral nifurtimox (10 mg/kg/day) was given three times a day for 10 days with eflornithine (400 mg/kg/day) infusions twice daily for 7 days. Success was assessed at 18 months, patient being alive, having no evidence of trypanosomes in any fluid, not requiring rescue medication, and CSF leukocyte count ≤20 cells per μl. The success rate of FNZ was lower than the combination (91% vs 98%), but the difference was within the predetermined acceptability margin of 13% [16]. In the subgroup of patients with severe CNS disease, >100 leucocytes in the CSF, FNZ showed less favorable outcome compared to NECT (86.9% vs 98.7%), but in the less severe cases (CSF leucocytes <100) treatment success was 98.7% for FNZ and 95.9% for NECT. The rates of death and treatment-related serious adverse event were similar, but some adverse events such as headaches, insomnia, and anxiety were more frequent with FNZ. This is the first oral monotherapy shown to be effective for the second stage HAT. This allowed home-based treatment and is more cost-effective to the healthcare system. It is the first oral treatment for all stages of *T. b. gambiense* HAT.

In an open-label, prospective, multicenter, cohort study, 227 patients with stage 1 gambiense HAT and 41 with early stage 2 (absence in CSF) were treated with FNZ [17]. Treatment was effective at 12 months in 227 of 230 patients (99%). No new safety issues were recognized in this study, and the commonest adverse events were headache and vomiting, and serious emergent adverse events were reported in 22 (9%).

### *11.3.5 Side Effects of Fexinidazole*

Although FNZ appeared to be mutagenic in the AMES test, no genotoxic potential was found from a series of in vitro, in vivo, or ex vivo tests in mammalian cells (Drug manual). There is no data on use in pregnant or lactating women and effects on embryo-fetal development in animals were secondary to maternal toxicity.

The most frequently reported adverse events from pooled data were vomiting (38%), nausea (33%), asthenia (20%), decreased appetite (17%), headache (16%), insomnia (15%), tremor (14%), and dizziness (14%) (see Drug manual). Psychiatric-related events (>1%) include hallucination and psychotic disorder. QTc prolongation of >450 ms have been seen in 7.2% of patients, but tachycardia is very uncommon.

## 11.4 Moxidectin

### *11.4.1 Onchocerciasis*

Onchocerciasis, also known as river blindness, is the second commonest cause of blindness due to an infection after trachoma. It is due to the filarial nematode, *Onchocerca volvulus,* transmitted by the bite of the female black fly (genus *Simulium damnosum*) that breeds along fast flowing rivers of parts of Africa and South America. River blindness is endemic in 31 countries of sub-Saharan Africa, the Arabian Peninsula, and the Amazonian focus straddling Venezuela and Brazil [18]. The WHO estimates that there are 25 million people with the disease and 90% are in sub-Saharan Africa.

Black flies inoculate patients with the infective microfilariae larvae during a blood meal and the microfilariae would develop into adult worms in 6–12 months that reside subcutaneously in fibrous nodules, commonly located around the head, torso, and the iliac crest. The fertilized adult worms produce 1000 to 1500 microfilariae daily and live for 10–15 years [18]. Dermatological manifestations are the commonest presentation, preceding ocular disease by years. These include severe pruritus, papular dermatitis, licheniform dermatitis, and areas of depigmentation, commonly around the shins—called "leopard skin." Ocular involvement is a result of inflammatory response to dead or dying microfilariae seeded to the eye. Vision loss can result from sclerosing keratitis, uveitis, optic nerve and chorioretinal atrophies, secondary cataract, and rarely glaucoma.

### 11.4.2 Treatment of Onchocerciasis

Diagnosis of onchocerciasis is usually made by examination of skin snips and slit-lamp examination of the eye and serological tests, enzyme-linked immunosorbent (ELISA), and Western blot. Diethylcarbamazine was first used for the disease, which was effective but caused serious ocular and systemic complications and is no longer recommended. Ivermectin was considered the treatment of choice up to recently; it kills the microfilariae but had no effect on the adult worm. The usual dose is 150 μg/kg orally once or twice a year for 10–15 years for the lifespan of the adult worm. Mass drug administration of ivermectin to endemic villages has been found to be effective to control the disease. The lack of safe and effective drugs to kill the adult worms has been a major restriction in the therapy of onchocerciasis.

The adult worm depends on an endosymbiotic bacteria, *Wolbachia*, for its survival and embryogenesis. Thus, therapy aimed at the bacteria can be used to kill the adult worm using doxycycline 200 mg daily for 6 weeks and ivermectin started 1 week before [19]. Doxycycline can kill 60% of adult female worms and sterilize up to 90%. However, this has not been instituted as standard therapy due to lack of robust data showing clinical ocular benefit. The efficacy of doxycycline with ivermectin in onchocerciasis was previously reviewed in the Cochrane Database [20]. Three randomized controlled trials included 466 patients treated with doxycycline and ivermectin versus ivermectin alone. However, the studies were considered low quality with missing data and selective outcome reporting.

Moxidectin was approved for treatment for human onchocerciasis by the US FDA in 2018, but it was used for years in veterinary medicine to treat farm animals with helminth infections. It is a semisynthetic macrocyclic lactone of the milbemycin class, a derivative of nemadectin which is a fermentation product of *Streptomyces* species [21].

### 11.4.3 Pharmacology of Moxidectin

Moxidectin is rapidly absorbed with peak plasma concentration in 3–4 h, and the total $AUC_{0-24}$ increases linearly with the dose [21]. It can be taken with or without food and a liquid formulation only increases the bioavailability modestly. Administration with a high fat meal increases the $C_{max}$ and AUC by 34–39% but is considered not clinically meaningful. It is lipophilic and may be retained in adipose tissue, widely distributed with a large volume of distribution (1.2 L/kg) and half-life of 20–43 days. The plasma protein binding is unknown in humans. Moxidectin is minimally metabolized in the body, it is not a substrate or inhibitor of CYP enzymes, renal elimination of the intact drug is negligible, and only 2% of the dose is excreted unchanged in the feces within 72 h. Mild to moderate renal impairment is not expected to affect the drug exposure, and it has not been studied in severe renal impairment. There is no known drug-drug interactions. A single dose of 8 mg is recommended for onchocerciasis.

### 11.4.4 Mechanism of Action of Moxidectin

The mechanisms of action of moxidectin have not been studied in *O. volvulus* but in other nematodes. It binds to glutamate-gated chloride ion channels vital to the function of invertebrate nerves and muscle cells, leading to influx of chloride ions, hyperpolarization, and muscle paralysis leading to death [21]. The drug is active against the microfilariae of *O. volvulus* and not effective in killing the adult worms, but it inhibits the intra-uterine embryogenesis and release of microfilariae from the adult worms (Moxidectin accessdata.fda.gov.).

Resistance of *O. volvulus* to moxidectin has not been found as yet but likely will occur with widespread use as it is common in nematodes of livestock. Continued use can result in therapeutic failure and cross-resistance between moxidectin and ivermectin [21].

### 11.4.5 Clinical Efficacy of Moxidectin

Phase II trial compared 2, 4, and 8 mg dose of moxidectin to the standard dose of ivermectin (150 μg/kg) and all three moxidectin doses resulted in faster and more complete clearance of microfilariae from skin than ivermectin [21]. Microfilaridermia reappeared after 2 months with ivermectin and progressively longer with increasing dose of moxidectin—up to 12 months after 8 mg, which was chosen for the dose in the phase III trial.

The phase III trial enrolled patients from 4 sites in sub-Saharan Africa to receive single oral dose of 8 mg moxidectin ($n$ = 978) or oral ivermectin at 150 μg/kg ($n$ = 494) and assessed clinical response at 1 year for skin microfilarial density [22]. The microfilarial density was significantly lower in the moxidectin group, treatment difference 86% ($p < 0.0001$). Mazzotti reactions (host response to dying microfilariae) occurred in 99% of moxidectin-treated patients and 97% of the ivermectin group, but no serious adverse reactions were related to treatment. Severe transient postural hypotension (causing dizziness and weakness) was the only efficacy-related grade 4 reaction that was more common in the moxidectin than the ivermectin group (5% vs 1%). Drug adverse events were rare with moxidectin, other than the Mazzotti reactions. Ocular Mazzotti reactions were similar in the two groups, 12% for moxidectin and 10% for ivermectin, including pruritus, conjunctivitis, eye pain, eyelid swelling, blurred vision, or tearing.

### *11.4.6 Adverse Effects of Moxidectin*

The adverse reactions to moxidectin, similar to ivermectin, were nearly all related to the Mazzotti reaction of the host to the dying microfilariae, and these include flu-like illness, pruritus, rash, musculoskeletal pain, headaches, ocular reactions, dizziness, weakness, enteritis, edema, worsening onchodermatitis, severe eosinophilia, leukocytosis, leucopenia, and increased liver enzymes (Moxidectin accessdata.fda.gov.). Similar to ivermectin, encephalopathy can occur with co-infection with *Loa loa*, which should be excluded before its use.

## 11.5 Tafenoquine

### *11.5.1* Plasmodium vivax *Malaria*

*Plasmodium vivax* malaria is the commonest form of malaria globally [23] and since 1980, relapse after treatment with chloroquine was thought to be due to activation of hypnozoites that reside in the liver. However, reactivation of malaria hypnozoites have only been shown recently [24]. Hence, standard treatment usually consists of 3-days chloroquine to clear the circulating trophozoites followed by 14 days primaquine to eradicate hypnozoites from the liver. For many decades, primaquine was the only drug available for radical cure of *P. vivax* and *P. ovale* malaria. Moreover, 14 days of primaquine often leads to poor compliance and incomplete treatment, and it can cause hemolytic anemia in subjects with glucose-6-phosphate dehydrogenase (G6PD) deficiency. However, the burden of *P. vivax* malaria decreased by 41.6% from 24.5 million cases in 2000 to 14.3 million cases in 2017 [23].

### 11.5.2 Pharmacology of Tafenoquine

Tafenoquine (TQ) was approved by the FDA for the radical cure of *P. vivax* malaria in 2018, the first new treatment in about 60 years. It is an 8-aminoquinoline derivative of primaquine which varies only by the presence of a 5-phenoxy group [25].

Oral TQ is slowly absorbed with peak plasma concentration in 8–12 h, and fatty meal increases the absorption by 30–40% [26]. The terminal plasma half-life is 12–16 days, and it is highly protein bound, 99.5%, but widely distributed with volume of distribution >200 liters. After degradation by several different pathways, TQ is slowly excreted primarily in the feces and renal elimination of the unchanged form is very low. The active metabolite 5,6 ortho-quinone-tafenoquine is produced in the liver by the activity of CYUP 2D6 microsomal enzyme (DrugBank Online). The dose of TQ for radical cure of vivax malaria is a single oral dose of 300 mg.

### 11.5.3 Antimalarial Activity of Tafenoquine

The mechanism of action of TQ is not fully known, but it appears that the active metabolite is internalized by the parasite and the oxidized metabolite produced hydrogen peroxide and hydroxyl radicals that lead to death of the parasite (Tafenoquine, DrugBank Online). It is active against the developing and dormant liver stages of the hypnozoites, the erythrocytes asexual stages (schizonticidal), gametocytes transmitted to mosquitoes, and sporozoites injected into humans by the mosquitoes.

### 11.5.4 Clinical Efficacy of Tafenoquine

Three randomized clinical trials in 9 malaria-endemic countries were conducted in 747 adult patients with *P. vivax* malaria [27]. All patients received chloroquine to clear the blood parasites and some groups received placebo, or primaquine 15 mg daily for 14 days, or single 300 mg dose of TQ. Patients receiving TQ had fewer relapse than those receiving placebo within 6 months and appeared to be similarly effective as primaquine. However, relapse of vivax malaria can occur up to a year which would not be captured by these studies. Overall the adverse events of TQ were similar to primaquine. TQ was found to have low efficacy to prevent vivax malaria in Indonesia when combined with dihydroartemisinin-piperaquine for unknown reasons [28].

TQ has also been studied for malaria prophylaxis in residents of malaria-endemic countries of Africa in placebo controlled studies and with mefloquine [29]. The drug was given as 3 daily loading doses plus weekly maintenance of 200 mg weekly for 10 weeks or more. Its pooled prophylactic efficacy was 93% and was similar to that of mefloquine 250 mg weekly.

### 11.5.5 *Safety of Tafenoquine*

TQ safety profile is similar to primaquine and patients before treatment should be tested for glucose-6-phosphate dehydrogenase (G6PD). Similar to primaquine, it can cause anemia, methemoglobinemia, leucopenia, and hemolytic anemia in subjects with G6PD deficiency (fda.gov/media/114755/download). In most studies, adverse events have been mild, most commonly nausea and abdominal cramps (≥5%) and CNS side effects (headaches, dizziness, abnormal dreams, insomnia, anxiety, and depression) [28]. TQ is not indicated in pregnancy, and it should be used with caution in psychotic disorders. Although primaquine can cause cardiac arrhythmia and prolongation of QT interval, no similar effects have been reported with TQ even in patients receiving the drug for 26 weeks [30].

## 11.6 Triclabendazole

### 11.6.1 *Human Fascioliasis*

Fascioliasis is a foodborne nematode infection, considered a neglected zoonosis by the WHO that has been reported in 81 countries with estimate of 2.6 million people infected globally [31]. The burden of disease of liver flukes is believed to be underestimated, and the disease appears to be emerging with climate change and 91 million people are at risk for infection globally.

Two trematode parasites are responsible for disease, *Fasciola hepatica,* reported throughout the world in patchy distribution, and *Fasciola gigantica,* present mainly in tropical regions of Africa, South and East Asia, and the Middle East. Sheep and cattle are the natural hosts of the parasites, but a wide range of wild and domestic mammals (46 species) can be infected, including humans [32]. Eggs shed in the stool of definitive mammal host embryonate in fresh water, releasing miracidia which infect 30 snail species (intermediate host), which release cercariae in the water that encyst to metacercariae on leafy water vegetables [32]. Humans become infected by eating contaminated watercress or other water plants. After ingestion, the metacercariae excyst in the intestines, allowing the immature parasites to penetrate the wall of the intestine, migrate in the abdominal cavity, and penetrate the liver to reach the bile ducts [32].

The acute migratory phase may last >12 weeks and associated with upper abdominal pain, fever, high eosinophilia, and hypodense track-like lesions on liver imaging. Mature parasites in the bile ducts produce inflammation, biliary obstruction, intermittent pain, and jaundice, which may result in subcapsular hematoma, liver abscess, liver fibrosis, and cirrhosis [32]. Otherwise asymptomatic patients may present with weight loss and anemia.

Livestock industry is markedly affected with estimates of 10–80% of dairy and meat cattle infected in developing and developed countries, and human populations in developing countries are mostly affected [32]. People living in poverty in small communities in the Andes Mountains of Bolivia, Ecuador, and Peru account for large portion of the global burden of fascioliasis. Other endemic countries include the Middle East (Egypt, Turkey, Iran), African countries, Asia (China and Vietnam), and some European countries (Portugal, Spain, and France).

Diagnosis of fascioliasis can be challenging and include abdominal imaging, ultrasound or computerized tomography (CT), serology and detection of the *Fasciola* spp. eggs, or antigen in duodenal aspirate or stools. Treatment includes anthelmintics to kill the flukes and symptomatic therapy.

### 11.6.2 Triclabendazole for Fascioliasis

Triclabendazole (TCZ) is a benzamidine derivative with a chlorinated benzene ring but no carbamate group, unlike other benzimidazole family of anthelmintics, that was developed and marketed by Ciba Pharmaceuticals (Fasinex) to treat fascioliasis in livestock in 1983, subsequently marketed for human use in the 1990s, and only approved by the FDA in 2019 for human fascioliasis [33].

TCZ, distributed by Novartis as Egaten for human fascioliasis, is a narrow-spectrum anthelmintic with activity only against *Fasciola* (*F. hepatica and F. gigantica*) and *Paragonimus* spp. [33]. The mechanism of action is not completely clear, but it may involve multiple targets, such as tegument disruption by preventing the polymerization of microtubules or adenylate cyclase activity [34]. The sulfoxide metabolite appears to have greater effect on the parasite motility than TCZ, through marked disruption of the integument and inhibition of protein synthesis [35].

### 11.6.3 Pharmacology of Triclabendazole

Oral TCZ is rapidly absorbed to produce mean $C_{max}$ in 3–4 h and food enhance the absorption to increase $C_{mx}$ and AUC of TCZ and sulfoxide metabolite two to three fold [33]. The drug is metabolized primarily by CYP1A2 and CYP2C9 into the active sulfoxide and sulfone metabolites, and the elimination half-life of TCZ and the metabolites is about 8 and 11–14 h [36]. The protein binding of TCZ was 96.7% and the metabolites 98.4–98.8% and the volume of distribution for the sulfoxide metabolite 1 L/kg. There is no excretion data in humans, but in animals the drug and the metabolites are largely excreted by the biliary tract and the feces (90%) with less than 10% in the urine [33]; its metabolites have the potential to inhibit many cytochrome P enzymes, the greatest inhibition on CYP2C19.

The recommended dose for acute or chronic fascioliasis is two doses of 10 mg/kg per dose with food separated by 12–24 h.

### 11.6.4 Clinical Efficacy of Triclabendazole

Clinical trials conducted over the past 25 years, and multiple case reports, found TCZ to have high efficacy (>70–100%, with dose dependent response) and was well tolerated [33]. Two randomized, controlled trials have been conducted. In Vietnam, 100 patients with fascioliasis were randomized to TCZ (2 doses of 10 mg/kg, 12 h apart) or artesunate 4 mg/kg once daily for 10 days [37]. The clinical response rate at 3 months (resolution of symptoms) was higher with TCZ (92%) versus 76% for artesunate ($p = 0.05$), but the improvement on imaging with ultrasound were similar (76% vs 70%). However, stools were not examined for microbiological response.

Uncontrolled studies that used fecal egg counts to assess efficacy of TCZ were summarized in a review article [33]. Different doses were used in various studies (10 mg/kg, 15 mg/kg and 20 mg/kg total dose) in 364 patients in 6 studies. The rate of clearance from the stools varied from 69% to 100% with the highest response usually occurring with 20 mg/kg given as two separate doses.

### 11.6.5 Treatment Failure and Resistance to Triclabendazole

Resistance of *F. hepatica* to TCZ has become widespread in livestock and has been reported in 17 endemic countries around the world, but the mechanisms of resistance are unknown [38]. Treatment failure and resistance of human cases of fascioliasis are sporadic and include a farmer in the Netherlands, four cases from Chile, one case from Turkey, and seven cases from Peru [39]. This is an emerging problem in the Andes of Peru, where 7 of 19 selected cases failed to clear *Fasciola* eggs after multiple courses of 2 doses of TCZ at 10 mg/kg per dose [40].

The underlying biochemical mechanisms of TCZ resistance remains unclear, but there is evidence of metabolic differences between susceptible and resistant isolates, indicating that altered uptake, efflux, and metabolism of TCZ are more important in the resistance than the tubulin-based process [39]. A single amino acid substitution in glutathione S-transfer in a resistant *F. hepatica* isolate had been detected, but this has not been confirmed in other resistant isolates [39].

### 11.6.6 Safety of Triclabendazole

TCZ is generally well tolerated and adverse reactions such as abdominal pain (56–93%), nausea (8–18%), and liver enzyme elevations (3.6–8%) may be due to expulsion of the liver fluke than from toxicity of the drug [36]. Other side effects greater than 2% include vomiting (6–7%), urticaria (7–11%), headache (6–14%), and musculoskeletal pain (4%).

**Table 11.1** Profile of recently approved anti-parasitic agents

| Agent | Disease | Parasite | Dose | Efficacy | Side effects |
|---|---|---|---|---|---|
| Benznidazole (old) | Chagas | *T. cruzi* | 5–8 mg/kg/day X 60 days | 60% (acute phase) | GI effects, rashes |
| Fexinidazole (new) | Sleeping sickness (HAT) | *T. b. gambiense* | 1800 mg X 4 & 1200 mg X 6 days | 86.6–91% | GI & CNS effects |
| Moxidectin (new) | River blindness | *O. volvulus* | 8 mg X 1 | 98% | Flu-like, rash, & Mazzotti reaction |
| Tafenoquine (new) | Malaria | *P. vivax and ovale* | 300 mg X 1 | 62–89% | GI & CNS effects, anemia |
| Triclabendazole (old) | Liver fluke (onchocerciasis) | F. *hepatica* & *F. gigantica* | 10 mg/kg X 2 | 69–100% | GI effects, rashes, and headaches |

Abbreviations: *CNS* central nervous system, *GI* gastrointestinal, *HAT* human African trypanosomiasis

## 11.7 Conclusion

Of the five drugs recently approved by the US FDA for treating invasive parasites, only three are actually new agents. For summary of the drugs profile, see Table 11.1. FNZ and moxidectin for African trypanosomiasis and onchocerciasis, respectively, represent real advance in the management of two serious neglected tropical diseases. TQ, a new drug for radical cure of *P. vivax and P. ovale*, represent a modest advance in the management of these malaria species, allowing single dose therapy which should improve compliance in completing treatment compared to 14 days of primaquine. BZN and TCZ are older drugs that were available in endemic countries of Chagas disease and fascioliasis, respectively, now allowing physicians in the US to access these drugs for infected migrants that moved to North America.

More effective and safer drugs are needed for serious and expanding parasitic infections such as multidrug-resistant falciparum malaria and the less frequent visceral leishmaniasis. Falciparum malaria resistant to artemisinin/artemether combinations are now spreading throughout Asia and eventually will be a global problem. Yet we have no new agents approved to meet this challenge. The time to act is now before it's too late.

## References

1. Pink R, Hudson A, Mouries M-A, Bendig M. Opportunities and challenges in antiparasitic drug discovery. Nat Rev. 2005;4:727–40.
2. WHO. Chagas disease [also known as American trypanosomiasis] World Health Organization, 1 April, 2021. https://www.who.int/news-room/fact-sheets/detail/chagas-disease-[american-trypanosomiasis

3. Manne-Goehle J, Umeh CA, Montgomery SP, Wirtz VJ. Estimating the burden of Chagas disease in the United States. PLoS Negl Trop Dis. 2016;10:e0005033.
4. Antinori S, Galimberti L, Bianco R, Grande R, Galli M, Coprbillino M. Chagas disease in Europe: a review for the internist in the globalized world. Eur J Intern Med. 2017;43:6–15.
5. Stanaway JD, Roth G. The burden of Chagas disease. Glob Heart. 2015;10:139–44.
6. Bern C, Montgomery SP, Herwaldt BL, et al. Evaluation of treatment of Chagas disease in the United States. JAMA. 2007;298:2171–81.
7. Perez-molina JA, Perez-Ayala A, Moreno S, Fernandez-Gonzalez C, Zamora J, Lopez-Valez R. Use of benznidazole to treat chronic Chagas' disease: a systematic review with a meta-analysis. J Antimicrob Chemother. 2009;64:1139–47.
8. Wilkinson SR, Kell JM. Trypanosomal drugs: mechanisms, resistance and new targets. Expert Rev Mol Med. 2009;11:e31.
9. de Higuchi ML, De Brito T, Martins Reis M, et al. Correlation between Trypanosoma cruzi parasitism and myocardial inflammatory infiltrate in human chronic chagasic myocarditis: light microscopy and immunohistochemical findings. Cardiovasc Pathol. 1993;2:101–6.
10. Morillo CA, Marin-Neto JA, Avezum A, et al. Randomized trial of benznidazole for chronic Chagas cardiomyopathy. NEJM. 2015;373:1295–306.
11. Hasslocher-Moreno AM, Saraiva RM, Sangenis LHC, et al. Benznidazole decreases the risk of Chagas disease progression and cardiovascular events: a long-term follow-up study. EClinicalMedicine. 2021;31:100694.
12. WHO. Trypanosomiasis, human African [sleeping sickness]. World Health Organization; 18 May 2021. https://www.who.int/news-room/fact-sheets/detail/trypanosomiasis-human-african-[sleepingsickness].
13. Torrele E, Trunz BB, Tweats D, Kaiser M, Brun R, Mazue G, Bray MA, Pecoul B. Fexinidazole—a new nitroimidazole drug candidate entering clinical development for the treatment of sleeping sickness. PLoS Negl Trop Dis. 2010;4:e923.
14. Kaiser M, Bray S, Cal M, Bopurdin Trunz B, Torreale E, Brun R. Antitrypanosomal activity of fexinidazole, a new nitroimidazole entering clinical development for the treatment of sleeping sickness. Antimicrob Agents Chemother. 2011;55:5602–8.
15. Betu Ku Mesu VK, Kalonji MK, Bardonneau C, et al. Oral fexinidazole for late-stage African *Trypanosoma brucei gambiense* trypanosomiasis: a pivotal multicenter, randomized, non-inferiority trial. Lancet. 2018;391:144–54.
16. Chappuis F. Oral fexinidazole for human African trypanosomiasis. Lancet. 2018;391:100–2.
17. Betu Ku Mesu VK, Kalonji WM, Bardonneau C, et al. Oral fexinidazole for stage 1 or early stage 2 African *trypanosome brucei gambiense* trypanosomiasis: a prospective, multicenter, open-label, cohort study. Lancet Glob Health. 2021;9:e999–1008.
18. Gyasi M, Okonkwo ON, Tripathy K. Onchocerciasis. NCBI Bookshelf. Updated Sept. 2021. https://www.ncbi.nim.nih.gov/books/NBK559027/
19. Hoerauf A, Mand S, Volkmann L, Buttner M, Marfo-Debrekyei Y, Taylor M, Adjei O, Buttner DW. Doxycycline in the treatment of human onchocerciasis: kinetics of Wolbachia endobacteria reduction and inhibition of embryogenesis in female Onchocerca worms. Microbes Infect. 2003;5:261–73.
20. Abegunde AT, Ahuja RM, Okafor NJ. Doxycycline plus ivermectin versus ivermectin alone for treatment of patients with onchocerciasis. Cochrane Database Syst Rev. 2016;(1):Art. No.: CD011146. https://doi.org/10.1002/1451858.CD011145.pub2.
21. Milton P, Hamley JID, Walker M, Basanez M-G. Moxidectin: an oral treatment for human onchocerciasis. Expert Rev Anti-infect Ther. 2020;18:1067. https://doi.org/10.1080/14787210.2020.1792772.
22. Opoku NO, Bakajika DK, Kanza EM, et al. Single dose moxidectin versus ivermectin for *Onchocerca volvulus* infection in Ghana, Liberia, and The Democratic Republic of the Congo: a randomized, controlled, double-blind phase 3 trial. Lancet. 2018;392:1207.
23. Battle KE, Lucas TCD, Nguyen M, et al. Mapping the global endemicity and clinical burden of *Plasmodium vivax*, 2000–17: a spatial and temporal modelling study. Lancet. 2019;394:332–43.

24. Markus MB. Transition from plasmodial hypnozoite to schizont demonstrated. Trends Parasitol. 2020;36:407–8.
25. Luk K-Y, Derbyshire ER. Tafenoquine: a step towards malaria elimination. Biochemistry. 2020;59:911–20.
26. Edstein MD, Kocisko DA, Brewer TG, et al. Population pharmacokinetics of the new antimalarial agent tafenoquine in Thai soldiers. Br J Clin Pharmacol. 2001;52:663–70.
27. Rodrigo C, Rajapakse S, Fernando D. Tafenoquine for preventing relapse in people with *Plasmodium vivax* malaria. Cochrane Database Syst Rev. 2020; https://doi.org/10.1002/14651858.CD01458.pub3.
28. Markus MB. Safety and efficacy of tafenoquine for *Plasmodium vivax* malaria prophylaxis and radical cure: overview and perspectives. Ther Clin Risk Manag. 2021;17:989–99.
29. Dow GS, Liu J, Hetzell B, et al. Summary of anti-malarial prophylactic efficacy of tafenoquine from three placebo-controlled studies of residents of malaria-endemic countries. Malar J. 2015;14. https://malariajournal.biomedicentral.com/articles/10.1186/s12936-015-0991-x
30. Chu CS, Hwang J. Tafenoquine: a toxicity overview. Expert Opin Drug Safety. 2021;20:349–62.
31. Furst T, Keiser J, Utzinger J. Global burden of human foodborne trematodiasis: a systematic review and meta-analysis. Lancet Infect Dis. 2021;12:210–21.
32. Carvedo MJ, Cabada MM. Human fasciolosis: current epidemiological status and strategies for diagnosis, treatment, and control. Res Reports Trop Med. 2020;11:149–58.
33. Gandhi P, Schmitt EK, Chen C-W, Samantray S, Venishetty VK, Hughes D. Triclabendazole in the treatment of human fascioliasis: a review. Trans R Soc Trop Med Hyg. 2019;112:797–804.
34. Kelley JM, Elliott TP, Beddoe T, et al. Current threat of triclabendazole resistance in *Fasciola hepatica*. Trends Parasitol. 2016;32:458–69.
35. Keiser J, Engels D, Buscher G, Utzinger J. Triclabendazole for the treatment of fascioliasis and paragonimiasis. Expert Opin Investig Drugs. 2005;14:1513–26.
36. EGATEN™ (triclabendazole) tablets. Product Mono[1]graph. 2019. Novartis Pharmaceuticals Corporation, East Hanover, New Jersey. Available at: https://www.accessdata.fda.gov/drugsatfda_docs/label/2019/208711s000lbl.pdf. Accessed 24 Mar 2022.
37. Hien TT, Ttruong NT, Minh NH, et al. A randomized controlled pilot study of artesunate versus triclabendazole for human fascioliasis in Central Vietnam. Am Trop Med Hyg. 2008;78:388–92.
38. Fairweather I, Brennan GP, Hanna REB, Robinson MW, Skuce PJ. Drug resistance in liver flukes. Int J Paarsitol Drugs Drug Resist. 2020;12:12–39.
39. Kelley JM, Ellioott TP, Beddoe T, Anderson G, Skuce P, Spithill TW. Current threat of triclabendazole resistance in *Fasciola hepatica*. Trends Parasitol. 2016;32:458–69.
40. Cabada MM, Lopez M, Cruz M, Delgado J, Hill V, White C Jr. Treatment failure after multiple courses of triclabendazole among patients with fascioliasis in Cusco, Peru: a case series. PLoS Negl Trop Dis. 2016;10:e0004361.

# Part IV
# New Antiviral Agents

# Chapter 12
# New Antiretroviral Agents for HIV Infection

## 12.1 Introduction

The development of potent combination antiretroviral therapy (ART) has revolutionized the management of HIV (human immunodeficiency virus) and the acquired immunodeficiency syndrome (AIDS). Development of these drugs was the most remarkable achievement of medical therapeutics. Now HIV can be easily managed with a single pill containing two or three highly active antiretroviral agents, resulting in lifespan almost similar to normal people. The therapeutics of HIV infection has come a long way since the first antiretroviral agent (zidovudine) was introduced in 1987. There are now eight separate classes of ART: nucleoside reverse transcriptase inhibitors (NRTIs), non-nucleoside reverse transcriptase inhibitors (NNRTIs), protease inhibitors (PIs), integrase strand transfer inhibitors (INSTIs), and four entry inhibitors, fusion inhibitor, CCR5 antagonist, attachment inhibitor, and post-attachment inhibitor (see Table 12.1). There are now 14 stand-alone dual or triple combination tablets and one injectable dual combination that are available for control of HIV-infection (see Table 12.2). There are other dual combination tablets that can be used along with another antiretroviral (ARV) for HIV control or for pre- or post-exposure prophylaxis. This chapter reviews the new ARVs approved by the Food and Drug Administration (FDA) in the United States (US) since 2018.



I. W. Fong, *New Antimicrobials: For the Present and the Future*, Emerging Infectious Diseases of the 21st Century,
https://doi.org/10.1007/978-3-031-26078-0_12

**Table 12.1** FDA-approved HIV medications

| Drug Class | Generic Name | Brand Name | FDA Approval Date |
|---|---|---|---|
| *Nucleoside reverse transcriptase inhibitors* | | | |
| | Zidovudine | Retrovir | March 1987 |
| | Lamivudine | Epivir | Nov. 1995 |
| | Abacavir | Ziagen | Dec. 1998 |
| | Tenofovir DF | Viread | Oct. 2001 |
| | Emtricitabine | Emtriva | July 2003 |
| *Non-nucleoside reverse transcriptase inhibitor* | | | |
| | Nevirapine | Viramune | June 1996 |
| | Efavirenz | Sustiva | Sept. 1998 |
| | Etravirine | Intelence | Jan. 2008 |
| | Rilpivirine | Edurant | May 2011 |
| | Doravirine | Pifeltro | Aug. 2018 |
| *Protease inhibitors* | Saquinavir | Invirase | Dec. 1995 |
| | Ritonavir | Norvir | March 1996 |
| | Atazanavir | Reyataz | June 2003 |
| | Fosamprenavir | Lexiva | Oct. 2003 |
| | Tipranavir | Aptivus | June 2005 |
| | Darunavir | Prezista | June 2006 |
| *Fusion inhibitors* | Enfuvirtide | Fuzeon | March 2003 |
| *CCR5 inhibitors* | Maraviroc | Selzentry | Aug. 2007 |
| *Integrase strand transfer inhibitors* | Raltegravir | Isentress | Oct. 2007 |
| | Dolutegravir | Tivicay | Aug. 2013 |
| | Cabotegravir | Vocabria | Jan. 2021 |
| *Attachment inhibitors* | Fostemsavir | Rukobia | July 2020 |
| *Post-attachment inhibitors* | Ibalizumab-uiyk | Trogarzo | March 2018 |

Abbreviations: *AF* alafenamide, *DF* disoproxil fumarate, *FTC* emtricitabine

## 12.2 Bictegravir/Emtricitabine/Tenofovir Alafenamide (Biktarvy)

Bictegravir (BIC) is a new INSTI with a high genetic barrier to the development of HIV-1 resistance that was combined with two NRTIs (emtricitabine [FTC] and tenofovir alafenamide [TAF]) in a single tablet, marketed as Biktarvy and approved by the FDA in February 2018. BIC is more active than the first-generation INSTIs, elvitegravir and raltegravir, and usually retain activity against HIV-1 strains with single or multiple mutations to these earlier INSTIs [1]. It is highly active in vitro against all HIV-1 groups and subtypes with 50% effective concentration ($EC_{50}$) of 0.02–6.6 nmol/L [1], similar to the earlier second-generation INSTI, dolutegravir (DTG). However, BIC was ≥two-fold more active than DTG against 13 of 47 clinical HIV isolates with INSTI resistance [1]. The combination of BIC with FTC and TAF exhibits high synergistic activity against HIV isolates in vitro.

**Table 12.2** Combination HIV medications

| Complete combinations | Generic name | Brand name | Approval date |
|---|---|---|---|
| 1. | Efavirenz/lamivudine/tenofovir DF | Atripla | July 2006 |
| 2. | Abacavir/lamivudine/zidovudine | Trizivir | Nov. 2000 |
| 3. | Emtricitabine/rilpivirine/tenofovir DF | Complera | Aug. 2011 |
| 4. | Elvitegravir/cobicistat/FTC/tenofovir DF | Stribild | Aug. 2012 |
| 5. | Abacavir/dolutegravir/lamivudine | Triumeq | Aug. 2014 |
| 6. | Eltegravir/cobicistat/FTC/tenofovir AF | Genvoya | Nov. 2015 |
| 7. | Emtricitabine/rilpivirine/tenofovir AF | Odefsey | March 2016 |
| 8. | Dolutegravir/rilpivirine | Juluca | Nov. 2017 |
| 9. | Doravirine/lamivudine/tenofovir DF | Delstrigo | Aug. 2018 |
| 10. | Darunavir/cobicistat/FTC/tenofovir AF | Symtuza | July 2018 |
| 11. | Bictegravir/emtricitabine/tenofovir AF | Biktarvy | Feb. 2018 |
| 12. | Dolutegravir/lamivudine | Dovato | April 2019 |
| 13. | Cabotegravir/rilpivirine | Cabenuva | Jan. 2021 |
| *Incomplete combinations* | | | |
| i. | Zidovudine/lamivudine | Combivir | Sept. 1997 |
| ii. | Lopinavir/ritonavir | Kaletra | Sept. 2000 |
| iii. | Emtricitabine/tenofovir DF | Truvada | Aug. 2004 |
| iv | Emtricitabine/tenofovir AF | Descovy | April 2016 |

Resistance to BIC in cell culture studies suggest that at least two amino acid substitution pathways may be involved to confer resistance, R263K/M501 and S153F plus transient T661, but these substitutions may be associated with minimal reductions in susceptibility to BIC (2–3-fold) [2]. Certain combinations of resistant mutations that confer resistance to elvitegravir and raltegravir and reduced susceptibility to DTG reduce the susceptibility to BIC. Extensive cross-resistance between DTG and BIC was observed when G14S/Q148H was present in combination with other INSTI mutations including T97A and L74M [3]. Evidently, resistance to FTC (M184V/I) and TAF (K65R) will reduce response to Biktarvy.

## *12.2.1 Pharmacokinetics of Biktarvy (BIC/FTC/TAF)*

All components of Biktarvy (BIC/FTC/TAF) are readily absorbed after oral administration with or without food with maximum plasma concentration ($C_{max}$) occurring between 0.5 and 4 hours for all the components [4]. Once daily dosing resulted in mean trough concentration of BIC (2038–2576 ng/ml) that was 13–16-fold higher than the mean effective inhibitory concentration against wild-type virus (162 μg/ml or 361 nmol/L). The plasma protein binding for BIC is >99%, for TAF 80%, and for FTC <4%. The median terminal plasma half-life (T1/2) of BIC is 17 h, 10 h for FTC, and 0.51 h for TAF and 32 h for tenofovir phosphate (the active moiety) [5].

BIC and TAF undergo metabolism for elimination, and FTC undergoes limited metabolism (oxidation and glucuronidation) and is eliminated primarily by the kidneys, 86% in the urine, and about 14% in the feces. BIC is metabolized mainly by CYP3A and UGT1A1 with elimination in the feces (60%) and 35% in the urine (metabolites). TAF is metabolized in peripheral blood mononuclear cells (PBMCs) and carboxylesterase-1 in the liver, and the major metabolite undergoes phosphorylation to form the active moiety tenofovir diphosphate, which is excreted primarily by the kidneys [5]. No dose adjustment is needed for creatinine clearance ($CR_{CL}$) ≥30 ml/min, and Biktarvy is not recommended in patients with $CR_{CL}$ < 30 ml/min or with severe hepatic impairment (not studied in this population). Biktarvy contains BIC 50 mg, FTC 200 mg, and TAF 25 mg.

### *12.2.2 Drug Interactions with Biktarvy (BIC/FTC/TAF)*

Potent inducers and inhibitors of the enzymes CYP3A and UGT1A1 and the intestinal transporters P-gp should be avoided with use of Biktarvy. Inducers of P-gp may increase TAF absorption and inhibitors decrease the absorption. Drugs that induce CYP3A, UGT1A1, and/or P-gp are contraindicated for use with Biktarvy—rifampin, rifapentine, carbamazepine, oxcarbazepine, phenobarbital, phenytoin, and St. John's wort—or inhibitors: atazanavir, cobicistat, azithromycin, clarithromycin, and cyclosporine [5].

BIC is an inhibitor of OCT2 and MATE1 and drugs that are substrates of these transporters may get increased plasma concentrations with co-administration with Biktarvy: dofetilide and metformin, which should be avoided or dose adjusted (metformin). Oral supplements with magnesium, aluminum, iron, or calcium and sucralfate should not be co-administered with Biktarvy as they can cause chelation of BIC, as with other INSTIs.

### *12.2.3 Clinical Efficacy of Biktarvy (BIC/FTC/TAF)*

Biktarvy (BIC/FTC/TAF) has been assessed in treatment-naïve and or treatment-experienced adult patients with HIV-1 infection in five randomized, comparator-controlled, multicenter, phase 3, noninferiority double-blind or open-label trials.

In treatment-naïve patients with HIV load ≥500 copies/ml and no resistance to the NRTI study drugs, two phase 3 trials were conducted to compare Biktarvy to DTG/FTC/TAF [6] and DTG/lamivudine (3TC)/abacavir (ABC), brand name Triumeq [7]. Randomized patients were stratified by viral load, CD4 cell count, and geographical region. Biktarvy was noninferior to each of the DTG-based regimen with attainment of viral load <50 copies/ml at 48 weeks (primary endpoint), and improvement in CD4 count at 48 weeks was similar between the different treatment arms. Longer-term follow-up of patients from these trials at 96 weeks showed

Biktarvy remains noninferior to the two DTG-based combinations with respect to viral load <50 copies/ml and mean absolute CD4 cell count [5]. Patient reported symptoms of fatigue, nausea, vomiting, dizziness, light-headedness, and difficulty sleeping over 48 weeks were less with Biktarvy than DTG/3TC/ABC (Triumeq) combination [8].

The efficacy of Biktarvy was assessed in three phase 3 trials in adult patients with controlled HIV with viral loads <50 copies for >3 or > 6 months on a stable ART regimen with a dual NRTI and either a PI (boosted atazanavir or darunavir) or an INSTI (DTG or boosted elvitegravir), randomized to staying on their usual ART or switching to Biktarvy [8–10]. There were no differences in the outcome at 48 weeks between Biktarvy and the other ART regimens in suppressed viral loads <50 or < 20 copies or changes in CD4 counts. Again, patients on Biktarvy had less symptoms of anxiety, nausea/vomiting, depressed feeling, and poor quality sleep than those on DTG/3TC/ABC [10].

### 12.2.4 Safety of Biktarvy (BIC/FTC/TAF)

Phase 3 randomized trials over 48 weeks have shown that Biktarvy was well tolerated with side effects profile similar to DTG/FTC/TAF or boosted elvitegravir combination or PI combinations, but better tolerated than DTG/ABC/3TC fixed combination. There were no cases of proximal tubulopathy and compared to DTG-based combinations, the lipid changes after 96 weeks were not significantly different or clinically relevant, and bone mineral density changes were similar [5].

A post-marketing study on adverse events of Biktarvy reported overall rate of 8.9% (rash, dizziness, nausea/vomiting, diarrhea, loss of appetite, weight gain, and fatigue) with 4% discontinuing therapy because of side effects (seven rashes, one insomnia and loss of appetite, and one feeling unwell) [11]. In a pooled analysis of eight randomized trials, INSTIs were associated with more weight gain than PIs or NNRTIs combinations, with DTG and BIC associated with more weight gain than elvitegravir/cobicistat, and TAF was associated with more weight gain than other NRTIs [12].

### 12.2.5 Biktarvy Place in Management of HIV-1

Biktarvy (BIC/FTC/TAF) fixed combination is a first-line ART similar in efficacy to DTG-fixed combinations, but more suitable than DTG/ABC/3TC (Triumeq) for patients co-infected with hepatitis B virus, since it has two agents effective against this virus (FTC and TAF). It is more potent with higher resistant barrier than earlier INSTI combinations with elvitegravir or raltegravir. Theoretically, Biktarvy should be effective against HIV-1 strains resistant to FTC/3TC with solitary 184 V mutations but studies are needed to test this hypothesis.

In view of the global increase in obesity and the potential complications, it would be best to avoid Biktarvy or other INSTI combinations in patients who are obese or overweight if other ART combinations are not contraindicated.

## 12.3 Dovato (Dolutegravir/Lamivudine)

Over the past 2 decades, the paradigm for successful ART was considered use of three-drug regimens, but simplifying therapy to a two-drug regimen has the advantage of potentially less drug toxicities, less drug-drug interaction, and reduced cost.

Juluca (DTG/rilpivirine) was the first fixed dual combination approved as complete treatment for HIV infection, approved in the US in 2017, but it was indicated only for patients with suppressed viral load for at least 6 months on a stable triple ART combination. Dovato (DTG/lamivudine [3TC]) was the second dual fixed combination approved by the FDA in 2019 for complete treatment of HIV-1, even in naïve patients with viral load <500,000 copies with no genotypic resistance to its components. The single tablet contains 50 mg of DTG and 300 mg of 3TC.

DTG has been available in a triple fixed combination of DTG/ABC/3TC (Triumeq) since 2014, and it is more potent than elvitegravir and raltegravir, with a high barrier to resistance and limited cross-resistance to the first-generation INSTIs, attributed to slower rate of disassociation from the integrase enzyme [13]. 3TC is a NRTI that requires intracellular phosphorylation and conversion to an active metabolite with excellent antiviral activity and well tolerated [14].

## 12.4 Pharmacokinetics of Dolutegravir/Lamivudine (Dovato)

DTG plasma level peaks 2–3 hours after ingestion with increased absorption after fatty meals (33–66%); it is highly protein bound (99%), with median volume of distribution of 17 liter and elimination half-life of 14 hours [15]. It is excreted mainly in the feces (63%) and the urine (31%) and is metabolized by UGT1A1 (major pathway) and CYP3A (minor pathway) [15].

3TC is rapidly absorbed with peak plasma levels at 1 hour, bioavailability of 87%, protein binding of 36%, volume of distribution of 96 liters, and half-life of 13–19 hours [15]. It is not significantly metabolized and it is primarily excreted by the kidneys.

## 12.5 Drug-Drug Interaction of Dovato (DTG/3TC)

The absorption of Dovato can be impaired with co-administration of divalent and trivalent cations (iron, calcium, aluminum, calcium, and magnesium) and sorbitol-containing products, and it should be taken 2 hours before or 6 hours after these agents. Several drugs can induce CYP3A metabolism of DTG and lower the blood concentration (carbamazepine, phenytoin, phenobarbital, rifampin, and St John's wort) and should be avoided with Dovato [16]. A few drugs plasma concentration can be increased by co-administration of Dovato by inhibition of OCT2 or MATE1-mediated elimination by DTG (metformin and dofetilide), and dose adjustment may be needed.

## 12.6 Clinical Efficacy of Dovato (DTG/3TC)

The clinical efficacy of Dovato was established in two phase III clinical trials, GEMINI-1 and GEMINI-2, of the two-drug fixed combination (DTG/3TC) versus three-drug regimen of DTG/FTC/tenofovir disoproxil fumarate (TDF) in ART-naïve adults with HIV-1 [17, 18]. These were multicenter, double-blind, randomized, non-inferiority trials of adults, ART-naïve with HIV-1 RNA ≤500,000 copies/ml without mutations to NRTIs, NNRTIs, or PIs. Patients with stage 3 HIV disease (except those with cutaneous Kaposi sarcoma and CD4 cell count <200 cells/μL), with severe hepatic impairment, and who are pregnant or breast feeding were excluded.

Of the 1433 patients included in the primary analysis, 80% had baseline viral load ≤100,000 copies and only 8% had baseline CD4 cell count ≤200 cells, with median CD4 count of 462 cells/μL. The primary response (HIV-1 RNA <50 copies/ml) rate at 48 weeks were similar between the 2 groups, 91% for the two-drug group and 93% in the three-drug arm. There was a lower response rate in patients with baseline CD4 count <200 cells/μL, 79% in the two-drug group and 93% in the three-drug arm. There were less adverse events in the two-drug group (mainly nausea), 18% vs 24%, respectively [17].

Secondary analysis of sustained HIV-1 RNA <50 copies/ml at 96 weeks showed that Dovato was noninferior to the three-drug regimen, 86% in the two-drug group and 89.5% in the three-drug group maintained viral load <50 copies/ml. Lower rate of viral suppression was found in patients with baseline CD4 count ≤200 cells/μL in the two-drug group (68.3% vs 87.3%), respectively [18]. Lower rate of adverse events continued to be reported for the two-drug group versus the three-drug group, 19.6% vs 25%, respectively.

In treatment-experienced patients, Dovato was also shown to be just as effective in maintaining HIV-1 RNA <50 copies/ml compared to continuing TAF-based three- or four-drug regimen at 48 weeks in the TANGO study, a phase III, open-label, multicenter, noninferiority trial in adults with viral load <50 copies/ml for >6 months [19].

### 12.6.1 *Safety Issues of Dovato (DTG/3TC)*

Women of child-bearing age should be counseled on using contraception and the small risk of DTG causing fetal toxicity in early pregnancy, neural tube defects 0.30% compared to 0.10% in non-DTG regimens [20]. The most common side effects in phase III trials have been headaches, diarrhea, upper respiratory tract symptoms, nausea, fatigue, and insomnia. INSTIs in general have been reported to cause increased lipids and weight gain [12]. Neuropsychiatric side effects have also been reported with DTG, including exacerbations of depressive disorders with suicide ideation [21].

## 12.7 Conclusions on Dovato

Dovato (DTG/3TC) is two-drug fixed combination ART suitable for treatment of naïve patients with HIV-1 RNA <500,000 copies/ml with no resistance to 3TC or INSTI and is well tolerated. Compared to fixed three-drug combinations, it appears to have less side effects and may be more cost-effective. In the manufacturer's base case, DTG/3TC was associated with fewer costs and higher quality-adjusted life-years than all comparator regimens [22].

However, caution should be used in selecting this two-drug combination for patients with CD4 count <200 cells/μL as the data is not robust and suggest it may not be as effective as three-drug combinations. It probably should not be a first choice for overweight and obese patients as the INSTI have greater risk than other ART in causing weight gain. Dovato should not be used in patients coinfected with chronic hepatitis B virus.

## 12.8 Delstrigo (Doravirine/Lamivudine/Tenofovir Disoproxil Fumarate)

Doravirine (DOR) is a novel NNRTI for the treatment of HIV-1, active against wild-type virus, and most common NNRTI-resistant variants (K103N, Y181C, G190A, Kio3N/Y181C, and E138K) at clinically achievable concentration [23]. It was approved by the FDA in 2018 as a 100 mg tablet to be used in conjunction with other ARVs and subsequently as a fixed three-drug combination (DOR/3TC/TDF), brand name Delstrigo, in September 2019.

DOR is a pyridinone NNRTI with potent activity against a wide range of HIV-1 subtypes with 95% effective concentration ($IC_{95}$) of 20 nmol/L (8.5 ng/mL) against wild-type virus and common efavirenz resistance mutants, $IC_{95}$ 54 nM [24]. It has a unique resistant profile, as resistant mutants selected with DOR may remain susceptible to efavirenz (EFV) and rilpivirine, and vice versa [25]. In the phase 3 study comparing DOR- and EFV-based regimens, the emergence of DOR resistance at virological failure was low, seven (1.9%) patients, but six of these had resistance to EFV as well [26]. European data showed that DOR resistance was 1.4% ($n = 9764$) in treatment-naïve patients, but higher in those previously treated with NNRTIs, intermediate resistance 12.7% and high-level resistance 6.1% [24].

### *12.8.1 Pharmacokinetics of Doravirine/Lamivudine/Tenofovir DF*

DOR is rapidly absorbed with maximum peak plasma concentration in 1–4 hours ($t_{max}$), bioavailability of 64%, not affected by administration with food, with protein binding of 75% and volume of distribution of 60.5 L [27]. CYP3A oxidation is the primary mechanism of elimination, and only about 6% is excreted unchanged in the urine, and the terminal half-life is ~15 hours [27].

TDF $t_{max}$ is 2–3 hours with 40% bioavailability, protein binding of <10% and 80% excreted by the kidneys with plasma half-life of 14 hours and intracellular half-life of 150–180 hours [28].

Delstrigo (DOR/3TC/TDF) contains 100 mg DOR, 100 mg 3TC, and 300 mg TDF for once daily and can be used in mild to moderate liver impairment, but DOR has not been studied in severe hepatic impairment. It should not be used in patients with creatinine clearance (CrCl) <50 ml/min, mainly due to the nephrotoxic potential of TDF.

Drugs that are strong cytochrome P450 CYP3A enzyme inducers can decrease DOR plasma concentration and decrease effectiveness of Delstrigo and should not be co-administered: carbamazepine, oxcarbazepine, phenobarbital, phenytoin, enzalutamide, rifampin, rifapentine, mitotane, and St. John's wort.

### *12.8.2 Clinical Efficacy and Safety of Delstrigo (DOR/3TC/TDF)*

The FDA approval of Delstrigo for complete treatment of HIV-1 was based on the phase 3, randomized, multicenter, double-blind, active controlled trial, DRIVE-AHEAD [29]. Once daily Delstrigo was found to be as effective as EFV/FTC/TDF in 780 adults with HIV-1 naïve to ART, with sustained viral suppression (HIV-1 RNA <50 copies), 84% vs 81%, respectively, at week 48. Further assessment at 96 weeks showed similar efficacy, viral suppression 77.5% vs 73.6%, respectively, with low virological failures in both groups [30]. Neuropsychiatric adverse events and rash were less with Delstrigo and discontinuation of treatment due to adverse events was lower within the first 48 weeks, 3% and 6%, respectively. Increase in fasting low-density lipoprotein cholesterol (LDL) and non-high-density lipoprotein cholesterol (non-HDL) were elevated at 96 weeks in the EFV/TFC/TDF group but not in the Delstrigo group.

DOR with FTC/TDF or ABC/3TC was also found to have similar efficacy as ritonavir-boosted darunavir with the same NRTIs in the DRIVE-FORWARD trial [31].

Delstrigo has been well tolerated in the phase 3 trials, and the most common side effects (≥5%) were dizziness (7%), nausea (7%), headache (6%), abnormal dreams (5%), fatigue (6%), and diarrhea (5%). TDF in Delstrigo has low risk of renal impairment (<3%) and decreased bone density (<10%).

### 12.8.3 *Conclusion on Delstrigo (DOR/3TC/TDF)*

Delstrigo can be used as another first-line treatment for adults with HIV-1 infections without resistant mutations to the contents, and it is a welcome addition to the fixed combinations for complete treatment. It is safe and well tolerated with mostly mild side effects, and unlike other NNRTI fixed combinations with rilpivirine (Complera and Odefsey), it can be taken with or without food and co-administration with a proton pump inhibitor (PPI) is not contraindicated.

## 12.9 Cabotegravir and Cabenuva (Cabotegravir/Rilpivirine)

Cabotegravir (CAB) is the first injectable ARV introduced for HIV treatment and prevention. It is a second-generation INSTI, closely related to DTG and BIC, with high barrier to resistance and significant activity against HIV mutants highly resistant to elvitegravir and raltegravir [32].

CAB is a member of the pyridinecarboxylic acids, defined by a pyridine ring with a carboxylic acid group and the five-membered ring is less flexible than the six-membered ring of DTG and BIC. It is highly active against wild-type virus and most single-mutant variants, but in vitro studies suggest that CAB has a lower resistant barrier than DTG and BIC and selection of Q148R/K with secondary mutations can confer high level resistance to the entire class of INSTI [33].

### 12.9.1 *Pharmacokinetics of Cabotegravir/Rilpivirine*

CAB has dual formulation of oral tablets and injectable nanosuspension. The oral CAB is rapidly absorbed with peak serum concentration of 2–3 hours after administration and half-life of about 40 hours [34]. Fifty-eight percent of the oral dose is excreted unchanged in the feces and 26.8% is mainly metabolized by uridine diphosphate glucuronosyltransferase (UGT); urinary excretion of the unchanged drug is <1%, but 27% of the glucuronide metabolite is excreted in the urine [34].

The long-acting injectable CAB (CAB-LA) by intramuscular injection (IM) produces peak serum concentration in 1 week with half-life between 21 and 50 days, due to the poor solubility of the nanoparticles in tissue [35]. It is highly protein bound (>99%) and the elimination is the same as the oral formulation. Subcutaneous injection (SC) results in similar maximum concentration and plasma concentration-time profile but causes more injection site reaction than the IM route [35]. The $IC_{90}$ of CAB is 166 ng/ml and monthly injections of 200 mg or 800 mg every 12 weeks resulted in concentrations 4 times the $IC_{90}$ [35].

No dose adjustment of CAB-LA is needed for mild to moderate hepatic impairment or severe renal impairment. Rifampin is the only medication contraindicated for co-administration as it can reduce the plasma concentration by 60% [34].

Rilpivirine (RPV), a NNRTI available in oral tablets or as fixed combination, was also formulated in nanocrystals to produce a long-acting injectable (RPV-LA) to be given together with CAB-LA for complete treatment of HIV-1, marketed as Cabenuva. The mean plasma concentrations of RPV-LA after a single injection during 28 days were comparable to the oral dosing [36]. RPV-LA has a half-life of 61–91 days and is highly protein bound (>99%); it is eliminated by the feces and undergoes metabolism via CYP450 3A4 [35].

Oral RPV is best taken with meals as the plasma levels are about 40% lower in the fasted state and suppression of gastric acid with a PPI also impairs absorption (Product monograph of Edurant [brand name]). Peak plasma concentration after oral administration is usually within 4–5 hours and the elimination terminal half-life is about 45 hours.

### *12.9.2 Clinical Efficacy of Cabenuva (CAB-LA/RPV-LA)*

Three multicenter, randomized, controlled, noninferiority, phase 3 trials have been conducted with Cabenuva (CAB/RPV): ATLAS, ATLAS-2 M, and FLAIR, after initial dose escalating phase 2 studies [35]. The endpoints were proportion of patients with HIV-1 RNA $\geq$50 copies ml and percent with RNA $\leq$50 copies at 48 weeks. ATLAS enrolled patients with undetectable viral load on conventional oral three-drug ART for at least 6 months (2 NRTI with NNRTI in 50%, or with INSTI in 33%, or with a PI in 17%), with 308 patients switching to Cabenuva IM every 4 weeks and 308 patients remaining on their oral ART for 52 weeks [37]. At week 48, the results were similar (noninferior) with 92.5% on long-acting injections (LAI) and 95.5% on oral ART maintaining viral load <50 copies/ml. Virological failure in each group was similar, three on LAI and four on oral ART. All three on LAI had NNRTI mutations (E138A, E138K/V1081, and E138E/K) and one had INSTI mutation (N155H). Patient satisfaction with therapy was greater in the LAI arm than the oral ART group.

The FLAIR trial enrolled HIV adults naïve to ART but were given induction with oral DTG/ABC/3TC (Triumeq) for 20 weeks, and after 16 weeks those with undetectable viral load were randomized to continue oral ART ($n$ = 283) or switch to Cabenuva IM ($n$ = 283) every 4 weeks for 100 weeks. The results met the criteria for noninferiority with maintenance of viral suppression in 93.6% versus 93.3% [38].

ATLAS-2 M enrolled patients with undetectable viral load to receive Cabenuva every 8 weeks ($n$ = 522) or every 4 weeks (523) for 100 weeks; at 48 weeks the results of patients with HIV-1 RNA <50 copies were similar and noninferior, 94.3% for every 8 weeks and 93.5% for every 4 weeks [39]. These results were maintained at week 96 analysis with 91% and 90%, respectively, maintaining undetectable virus; eight patients in the 2-monthly dosing and two in the 1-monthly dosing developed virological failure from the start of the study [40].

The LATTE-2 study followed patients in the randomized study of 2-monthly and 1-monthly LAI (CAB/RPV) for 5 years, with 186 of 230 (81%) maintaining viral load <50 copies at week 256 [41]. No virological failures occurred after 48 weeks, but three patients experienced serious adverse events with one fatal outcome.

### 12.9.3 Safety and Drug Interactions of Cabenuva (CAB/RPV)

The most common side effects of Cabenuva were injection site reactions (79%), pain, nodule formation, induration, and swelling with median duration of 3 days and 88% resolved within 7 days. Significant weight gain has been reported with INSTI and was noted in long-term follow-up of the LATTE trial, median weight-gain 3 kg at 96 weeks and 6.5 kg at 312 weeks [42]. Elevated liver enzymes three times the upper limit of normal in alanine aminotransferase (ALT) was found in >2% of patients in the phase 3 trials, but many of these were related to acute hepatitis A, B, and C [42]. Other side effects were fever (8%), fatigue (5%), headaches (4%), muscular pain (3%), nausea (3%), sleep disturbance (2%), dizziness (2%), rash (2%), and 4% discontinued CAB/RPV due to adverse events [42].

The US product monograph list several drugs that should not be co-administered with Cabenuva: carbamazepine, oxcarbazepine, phenobarbital, phenytoin, rifampin, rifabutin, rifapentine, dexamethasone, and St. John's wort (Cabenuva, monograph, Vivv Healthcare).

### 12.9.4 Indication and Dosing of Cabenuva (CAB/RPV)

Cabenuva was approved by the FDA for complete treatment of HIV-1 in January 2021 in patients virologically suppressed on oral ART with no resistant mutations to the contents. An oral lead-in of the drugs to assess tolerability is necessary before starting the LAI, 28 days of oral CAB 30 mg and oral RPV 25 mg once daily with meals. Then IM loading of two 3 ml injections of 600 mg CAB and 900 mg RPV, followed by two 2 ml injections of CAB 400 mg and RPV 600 mg monthly (monthly schedule). For the 2-monthly dosing, CAB 600 mg and RPV 900 mg are given monthly for the first 2 months, then 2-monthly thereafter.

### 12.9.5 Cabotegravir for Prevention of HIV

CAB-LAI (Apretude) was approved by the FDA in December 2021 for HIV pre-exposure prevention (PrEP). Apretude for the first tow injections is given 1 month apart and then every 2 months thereafter. Patients may either start their treatment with CAB-LAI or oral CAB (Vocabria) for 4 weeks to assess their tolerance to the medications.

Apretude was superior to oral PrEP with TDF/FTC (Truvada) in men who have sex with men (MSM) and transgender women in a randomized, double-blind trial [43]. Over 4500 participants were randomized to CAB-LAI 600 mg every 2 months compared to Truvada once daily for 153 weeks. CAB showed a 66% lower incidence of HIV infection, 13 incident infection vs 39 in the TDF/FTC group. INSTI resistance was found in five subjects on CAB, one with baseline infection and four with incident infection. Thirty-three percent of all participants reported grade 3 adverse events, and 5% reported serious adverse events including seizures and liver-related events leading to discontinuation of medications, similar between the groups. Injection site reaction occurred in 81% of participants, but only 2.4% discontinued due to such reactions.

Women in sub-Saharan Africa are one of the groups at highest risk of HIV in the world, and oral PrEP has been largely ineffective from low compliance. In a phase 3, randomized, controlled superiority trial comparing CAB-LAI every 8 weeks to daily oral TDF/FTC in 3224 women from 7 countries in sub-Saharan Africa, CAB injections resulted in 88% lower risk of HIV (4 in CAB arm vs 36 in TDF/FTC arm) [44]. Adverse events and discontinuation were similar between the two groups, but the study was stopped early due to efficacy with median follow-up of 1.24 years.

### 12.9.6 Conclusion on Cabotegravir and Cabenuva

CAB-LAI should be considered the PrEP of choice in African women for the prevention of HIV and should be offered as an option in MSM, especially black subjects with poor oral compliance. The concern, however, is decreased compliance over longer periods (3–5 years). Hence, longer-term observational data on compliance, efficacy, and INSTI resistance should be collected.

Cabenuva is a valuable addition to the ART arsenal for complete treatment of HIV in selected patients with difficulty swallowing pills and those with psychiatric conditions where supervised LAI are best for control. The data shows the LAI combination is effective with tolerable side effects.

## 12.10 Ibalizumab-uiyk (Trogarzo)

Ibalizumab (Trogarzo) is the first monoclonal antibody approved for the treatment of HIV-1 infection. It was approved by the FDA in March 2018 to be used in combination with other ARV for heavily treated patients with multidrug-resistant HIV-1 infection failing their current regimen.

The entry of HIV into cells (mainly CD4-lymphocytes) involves viral attachment, co-receptor binding, and fusion, with agents available for interrupting the last two steps. The fusion inhibitor enfuvirtide was the first entry inhibitor to be introduced in 2003, followed by maraviroc the CCR5 co-receptor antagonist in 2007. Ibalizumab (IBZ) is the first-in-class CD4-directed post-attachment inhibitor and the first monoclonal antibody approved for HIV treatment.

### *12.10.1 Pharmacodynamics and Antiviral Activity of Ibalizumab*

IBZ prevents HIV-1 entry in cells post-attachment by preventing conformational changes in the gp 120-CD4 complex that enable co-receptor binding and fusion [45]. It blocks HIV-1 from infecting $CD4^+$ T cells by binding to domain 2, preventing viral transmission that occurs via cell-cell fusion. IBZ also prevents HIV-1 induced syncytium formation between infected and uninfected CD4 cells [46]; syncytium formation has been associated with HIV pathogenesis and progression to AIDS.

Clinical and laboratory strains of CCR5- and CXC4-tropic HIV-1 in peripheral lymphocytes are inhibited by IBZ with median effective concentration of 50% ($EC_{50}$) of 8 ng/ml against HIV-1 group M isolates (subtypes A, B, C, D, E, and O) [46]. The antiviral activity is correlated with complete CD4 cell receptor coating with IBZ in the early dosing period; in phase 1 study, complete receptor coating was noted with serum concentration of IBZ >5 μg/ml [46]. IBZ retain antiviral activity against multidrug-resistant HIV-1 strains including isolates with enfuvirtide resistance.

### *12.10.2 Resistance to Ibalizumab*

The main mechanism of resistance to IBZ is by reduced expression or loss of potential N-linked glycosylation sites (PNGS) in the V5 loop of gp 120 [47]. Marked resistance to IBZ was associated with complete absence of PNGS in the N-terminus of V5 and the number of V5 PNGS was correlated with IBZ susceptibility. There is no evidence of cross-resistance between IBZ and other classes of ARVs.

### 12.10.3 Pharmacokinetics of Ibalizumab

Following infusion of IBZ over 0.5–1.5 hour of a single loading dose of 2000 mg followed by maintenance dose of 800 mg every 2 weeks (the recommended dose), steady-state concentrations (mean concentration > 30 μg/ml) are achieved after the first maintenance dose [46]. The median time to and $C_{max}$ is dose dependent, but the increases in AUC are greater than dose-proportional. The volume of distribution is approximately that of serum volume, 4.8 L, and the elimination half-life (37.8 h to 64.1 h) is dose dependent and non-linear [48]. Since IBZ is a protein, it is expected to be degraded into small peptides and amino acids.

No dosage adjustment is considered necessary for weight, and hepatic and renal impairments are not expected to impact on the pharmacokinetics. IBZ is not expected to have any drug-drug interactions.

### 12.10.4 Efficacy of Ibalizumab

Two phase II randomized, double-blind trials were conducted to assess the dosing and feasibility of utilization of IBZ in heavily treated patients with virological failure and resistance against >1 class of ARVs [46]. Based on the results, 2000 mg loading dose followed by 800 mg every 2 weeks was chosen for the phase III TMB-301 trial. TMB-301 was a multicenter, open-label, single-arm study in 40 heavily treatment-experienced HIV-infected subjects with multidrug-resistant HIV-1 with viral load >1000 copies/ml and resistance to at least 1 ARV from each class [49]. The patients were on ART for at least 6 months and failing therapy.

The patients received a IV loading dose of 2000 mg IBZ while continuing previous ART regimen and were initiated on an optimized background regimen (maintenance period day 14 to week 25) with IBZ 800 mg every 14 days, starting on day 21. The degree of viral resistance in the selected patients were as follows: 50% had HIV resistant to all drugs in ≥3 ARV classes, 33% had resistance to all drugs in 4 classes, and 13% had resistance to all approved ARVs. At week 25, 50% of patients had achieved a viral load of <200 copies/ml; 43% had a viral load <50 copies; and mean reduction of viral load in the cohort was 1.6 $\log_{10}$ copies/ml. The mean increase in CD4 count was 62 cells/μL with lower increase in those with baseline CD4 count <50 cells/μL than those with >50 cells/μL.

IBZ with an optimized backbone maintained viral suppression up to 96 weeks in patients from TMB-301 study who entered the expanded access program, which also included another cohort of 38 patients [46]. Decreased susceptibility to IBZ has been observed in patients with viral failure and may be associated with genotypic changes in the HIV-1 envelope coding sequence [48].

### *12.10.5 Safety of Ibalizumab*

Most of the adverse reactions with IBZ were mild to moderate, including diarrhea, dizziness, fatigue, nausea, fever, and rash [46]. Serious adverse events were reported in 28% in the TMB-301 trial with 13% discontinuing treatment due to these events. As with any ART in patients with AIDS, two patients receiving IBZ with backbone ARVs have been reported to develop immune reconstitution inflammatory syndrome (IRIS). Hypersensitivity reactions with rash to IBZ occurred in 1–3 weeks and were mild-moderate in severity and resolved within 1–3 weeks with continued administration.

### *12.10.6 Conclusion on Ibalizumab*

Although the number of patients treated with IBZ is limited, it is a useful addition to the arsenal of drugs to fight HIV and AIDS. Luckily, the number of patients with multidrug-resistant HIV infections are few and far between in any one HIV/AIDS treatment center. The main disadvantage of this agent is the need for IV infusion every 2 weeks. Looking to the future, the development of a new class of ARV that can be administered IM every 6 months, lenacapavir, is exciting. It works by interfering with assembly and disassembly of the HIV capsid and should be effective against multidrug-resistant HIV (Pebody R. Potent new anti-HIV drug only needs to be injected every 6 months. News in Brief, 26 July 2021). Studies with lenacapavir in combination with another new ARV, islatravir, have recently been started.

## 12.11 Fostemsavir

Fostemsavir (brand name Rukobia) is the latest ARV approved by the FDA for use in combination with other ARVs in July 2020. It is also approved by the European Medicines Agency (EMA) for the treatment of multidrug-resistant HIV-1 in heavily treated patients.

Fostemsavir (FTS) is a prodrug of temsavir that prevents the binding and entry of HIV-1. It is a first-in-class attachment inhibitor that binds directly to the viral envelope glycoprotein 120 (gp120) adjacent to the CD4 binding site [50]. Fostemsavir (FTS) is a methyl-phosphate prodrug with enhanced bioavailability due to increased solubility in the gastrointestinal tract. It has no in vitro cross-resistance with other classes of ARV, including entry inhibitors, and it can be used irrespective of HIV tropism [51]. The active drug (temsavir) demonstrated broad spectrum of activity against HIV-1 clinical isolates, with $IC_{50}$ ranging from subnanomolar concentrations to >0.1 μM [52]. Amino acid substitution at four positions in gp120 (S375H/

IN/M/T, M426L/P, M4341/K, M4341/K, and M4751) can affect the susceptibility of the virus to temsavir [53].

### 12.11.1 Pharmacokinetics of Fostemsavir

FTS is available as an extended-release 600 mg tablet administered orally twice daily. Its bioavailability is about 27%, not significantly affected by food, with a $T_{max}$ of about 2 hours, and it is rapidly hydrolyzed to the active metabolite (temsavir) by alkaline phosphatases in the brush border of the intestines [54]. Temsavir reaches steady state in 2–3 days with half-life of 11 hours, $C_{max}$ of 1770 ng/ml, protein binding of 88.4%, and volume of distribution of 29.9 liters (Rukobia[R] Summary of Product Characteristics, ViiV Healthcare). Temsavir is highly metabolized and the inactive metabolites are excreted in the urine and feces. About 36.1% of the dose is metabolized by esterases and 21.2% is metabolized by CYP3A4. Approximately 51% of the metabolites are excreted in the urine and 33% in the feces, and < 2% of the unchanged parent drug appears in the urine and 1.1% in the feces.

Co-administration of strong CYP3A inducers can decrease blood levels of temsavir and should be avoided (i.e., rifampin, carbamazepine, phenytoin, phenobarbital, and St. John's wort) [55].

### 12.11.2 Clinical Data on Fostemsavir

In phase 2b trial of moderately pretreated HIV adults, FTS monotherapy after 7 days resulted in median decrease of HIV-1 RNA by 0.69 $\log_{10}$ to 1.44 $\log_{10}$ copies/ml [56]. Combination of FTS with raltegravir/TDF for 192 weeks was found to be well tolerated with comparable efficacy as ritonavir-boosted atazanavir/raltegravir/TDF [57].

The efficacy of FTS in adults with multidrug-resistant HIV-1 infection was assessed in the phase III trial, BRIGHTE study, which enrolled 372 patients failing ART [58]. There were 272 participants with at least one fully active ARV in <2 classes entered in the randomized cohort and 99 without any remaining option who were assigned the nonrandomized cohort. In the randomized cohort, patients were randomized to FTS (600 mg twice daily) or placebo in combination with their failing regimen for 8 days, then open-label FTS with an optimized backbone. In the nonrandomized cohort, FTS at the same dose was started at the onset in combination with a regimen based on resistance testing and ART history. At day 8, the HIV-1 RNA decreased by a mean level of 0.79 $\log_{10}$ in the FTS group and 0.17 $\log_{10}$ in the placebo arm ($p < 0.001$). At week 48, 54% of the patients in the randomized cohort and 38% in the nonrandomized cohort achieved viral load <40 copies/ml; the mean increase in CD4 count was 139 cells/ml$^3$ and 64 cell/ml$^3$, respectively. FTS was discontinued in 7% of patients due to adverse events.

Virological failure occurred in 18% of the randomized cohort and 46% of the nonrandomized cohort. In the randomized cohort, gp120 substitutions were found in 20 of 47 patients (43%) with virological failure; S375N and M426L were the most frequent substitutions. While, in the nonrandomized cohort gp120 substitutions were found in 70% with a median change in the $IC_{50}$ of temsavir relative to the baseline was an increase by a factor of 470.

Week 96 results for BRIGHTE showed comparable rates of virological and immunologic responses (randomized cohort) and safety (combined cohorts) among subgroups [59]. Virological response rates (HIV-1 RNA <40 copies/ml) were mainly associated with the overall susceptibility of the new optimized background agents. Participants with baseline CD4 count <20 cells/µl had a mean increase of 240 cells/µL, and adverse events and deaths were greater in patients with baseline CD4 < 20 cells/µL than those with at least 200 cells/µL.

### 12.11.3 Safety of Fostemsavir

FTS was well tolerated in the phase 2 and 3 trials, with nausea, diarrhea, and headaches being the most common side effects, but others include rash, abdominal pain, insomnia, abnormal dreams, and sleepiness. Laboratory abnormalities that were absent in phase 2 studies but seen in the phase 3 trial include creatinine elevations (grade 3/4) in 17% and grade 3/4 biliary abnormalities in 24%, which may be due to concomitant medications [60]. Three patients (<1%) had QTc prolongation at the approved dose resulting in discontinuation of the drug [60].

### 12.11.4 Conclusion on Fostemsavir

FTS, an oral attachment inhibitor, is a welcome addition for patients with multidrug-resistant HIV infections. Although rates of highly resistant HIV-1 infection are decreasing in North America, rates of resistance are increasing in developing countries due to lack of genetic and viral load testing and pretreatment exposures. FTS is well tolerated and effective regardless of tropism but requires an optimized backbone to maintain efficacy.

The future prospect of managing multidrug-resistant HIV-1 appears bright, with the marketing of ibalizumab and fostemsavir and development of lenacapavir, the first-in-class capsid inhibitor, and islatravir, the first-in-class nucleoside reverse transcriptase translocation inhibitor [61, 62].

# References


1. Tsiang M, Jones GS, Goldsmith J, et al. Antiviral activity of bictegravir [GS-9883], a novel potent HIV-1 integrase strand inhibitor with an improved resistance profile. Antimicrob Agents Chemother. 2016;60:7086–97.
2. US FDA. Application number: 210251Orig1s000 multi-discipline review. 2017. https://www.accessdata.fda.gov/. Accessed 4 April 2022
3. Zhang WW, Cheung PK, Oliviera N, et al. Accumulation of multiple mutations in vivo confers cross-resistance to new and existing integrase inhibitors. J Infect Dis. 2018;218:1773–6.
4. Gilead Sciences. Biktarvy [bictegravir, emtricitabine, and tenofovir alafenamide]: US prescribing information. 2018. https://www.accessdata.fda.gov/. Accessed 6 April 2022
5. Deeks DDF. Bictegravir/emtricitabine/tenofovir alafenamide: a review in HIV-1 infection. Drugs. 2018;78:1817–28.
6. Sax PE, Pozniak A, Ml M, et al. Coformulated bictegravir, emtricitabine, and tenofovir alafenamide versus dolutegravir with emtricitabine and tenofovir alafenamide, for initial treatment of HIV-1 infections [GS-US-380-1490]: a randomized, double-blind, multicenter, phase 3, non-inferiority trial. Lancet. 2017;390:2073–82.
7. Gallant J, Lazzarin A, Mills A, et al. Bictegravir, emtricitabine, and tenofovir alafenamide versus dolutegravir, abacavir and lamivudine for initial treatment of HIV-1 infection [GS-US-380-1489]: a double-blind, multicenter, phase 3, randomized controlled non-inferiority trial. Lancet. 2017;390:2063–72.
8. Wohl D, Clarke A, Maggio F, et al. Patient-reported symptoms over 48 weeks among participants in randomized, double-blind, phase III non-inferiority trials of adults with HIV on co-formulated bictegravir, emtricitabine, and tenofovir alafenamide versus co-formulated abacavir, dolutegravir, and lamivudine. Patient. 2018;11:561–73.
9. Daar ES, De Jesus E, Ruane P, et al. Efficacy and safety of switching to fixed-dose bictegravir, emtricitabine, and tenofovir alafenamide from boosted protease inhibitor-based regimens in virologically suppressed adults with HIV-1: 48 week results of a randomized, open-label, multicenter, phase 3, non-inferiority trial. Lancet HIV. 2018;5:e347–e56.
10. Molina J-M, Ward D, Brar I, et al. Switching to fixed-dose bictegravir, emtricitabine, and tenofovir alafenamide from dolutegravir plus abacavir and lamivudine in virologically suppressed adults with HIV-1: 48 week results of a randomized, double-blind, multicenter, active-controlled, phase 3, non-inferiority trial. Lancet HIV. 2018;5:e357–e65.
11. Hyes E, Derrick C, Smalls D, Smith H, Kremer N, Weissman S. Adverse events with Biktarvy: post-marketing study. Open Forum Infect Dis. 2019;6(Suppl 2):S863–4.
12. Sax PE, Erlandson KM, Lake EL, et al. Weight gain following initiation of antiretroviral therapy: risk factors in randomized comparative clinical trials. Clin Infect Dis. 2020;71:1379–89.
13. Hightower KE, Wang R, Deanda F, et al. Dolutegravir [S/GSK 1349572] exhibits significantly slower dissociation than raltegravir and elvitegravir from wild-type and integrase-resistant HIV-1 integrase –DNA complexes. Antimicrob Agents Chemother. 2011;55:4552–9.
14. Quercia R, Perno CF, Koteff J, et al. Twenty-five years of lamivudine: current and future use for the treatment of HIV-1 infection. J Acquir Immune Defic Syndr. 2018;78:125–35.
15. Zamora FJ, Dowers E, Yasin F, Ogbuagu O. Dolutegravir and lamivudine combination for the treatment of HIV-1 infection. HIV AIDS [Auckl]. 2019;11:255–63.
16. Santevecchi BA, Miller S, Childs-Kean LM. Doing more with less: review of dolutegravir-lamivudine, a novel single-tablet regimen for antiretroviral naïve adults with HIV-1 infection. Ann Pharmacother. 2020;54:1252–9.
17. Cahn P, Madero JS, Arribas JR, et al. Dolutegravir plus lamivudine versus dolutegravir plus tenofovir disoproxil fumarate and emtricitabine in antiretroviral-naïve adults with HIV-1 infection [GEMINI-1 and GEMINI-2]: week 48 results from two multicenter, double-blind, randomized, non-inferiority, phase 3 trials. Lancet. 2019;393:143–55.

18. Cahn P, Madero JS, Arribas JR, et al. Durable efficacy of dolutegravir plus lamivudine in antiretroviral treatment-naïve adults with HIV-1 infection: 96- week results from the GEMINI-1 and GEMINI-2 randomized clinical trials. J Acquir Immune Defic Syndr. 2020;83:310–8.
19. Van Wyk J, Ajana F, Bisshop F, et al. Efficacy and safety of switching to dolutegravir/lamivudine fixed-dose 2-drug regimen vs continuing a tenofovir alafenamide-based 3- or 4-drug regimen for maintenance of viral suppression in adults living with human immunodeficiency virus type 1: phase 3, randomized, noninferiority TANGO study, Clin Infect Dis 2020; published online January 6. https://doi.org/10.1093/cid/ciz1243
20. Zash R, Holmes L, Diseko M, et al. Neural-tube defects and antiretroviral treatment regimens in Botswana. N Engl J Med. 2019;381:827–40.
21. Caihol J, Rouyer C, Alloui C, Jeantils V. Dolutegravir and neuropsychiatric adverse events: a continuing debate. AIDS. 2017;31:2023–4.
22. Canadian Agency for Drug and Technologies in Health. Pharmacoeconomic review report: dolutegravir/lamivudine [Dovato]. 2019; https://www.ncbi.nim.nih.gov/books/NBK551931/
23. Feng M, Sachs NA, Xu M, et al. Doravirine suppresses common nonnucleoside reverse transcriptase inhibitor-associated mutants at clinically relevant concentrations. Antimicrob Agents Chemother. 2016;60:2241–7.
24. Lai MT, Feng M, Falgueyret JP, et al. In vitro characterization of MK-1439, a novel HIV-1 nonnucleoside reverse transcriptase inhibitor. Antimicrob Agents Chemother. 2014;58:1652–63.
25. Feng M, Wang D, Grobler JA, Hazuda DF, Miller MD, Lai MT. In vitro resistance selection with doravirine [MK-1439]. A novel nonnucleoside reverse transcriptase inhibitor with distinct mutation development pathways. Antimicrob Agents Chemother. 2015;59:590–8.
26. Orkin C, Squires KE, Molina JM, et al. Doravirine/lamivudine/tenofovir disoproxil fumarate is non-inferior to efavirenz/emtricitabine/tenofovir disoproxil fumarate in treatment-naïve adults with human immunodeficiency virus-1 infection: week 48 results of the DRIVE-AHEAD trial. Clin Infect Dis. 2019;68:535–44.
27. Boyle A, Moss CE, Marzolini C, Khoo S. Clinical pharmacodynamics, pharmacokinetics, and drug interaction with profile of doravirine. Clin Pharmacokinet. 2019;58:1553–65.
28. Bazzoli C, Jullien V, Le Tiec C, Rey E, Mentre F, Taburet A-M. Intracellular pharmacokinetics of antiretroviral drugs in HIV-infected patients, and their correlation with drug action. Clin Pharmacokinet. 2010;49:17–45.
29. Orkin C, Squires KE, Molina J-M, et al. Doravirine/lamivudine/tenofovir disoproxil fumarate [TDF] versus efavirenz/emtricitabine/tenofovir disoproxil fumarate in treatment-naïve adults with human immunodeficiency virus type 1 infection: week 48 results of DRIVE-AHEAD trial. Clin Infect Dis. 2019;68:535–44.
30. Orkin C, Squires KE, Molina J-M, et al. Doravirine/lamivudine/tenofovir disoproxil fumarate [TDF] versus efavirenz/emtricitabine/TDF in treatment-naïve adults with human immunodeficiency virus type 1 infection: week 96 results of the randomized, double-blind, phase 3 DRIVE-AHEAD noninferiority trial. Clin Infect Dis. 2021;73:33–42.
31. Molina JM, Squires K, Sax PE, DRIVE-FORWARD Study Group, et al. Doravirine versus ritonavir-boosted darunavir in antiretroviral-naïve adults with HIV-1 [DRIVE-FORWARD]: 48-week result of a randomized, double-blind, phase 3, non-inferiority trial. Lancet HIV. 2018;5:e211–20.
32. Karmon SL, Mohri H, Spreen W, et al. GSK1265744 demonstrates robust activity in vitro against various clades of HIV-1. J Acquir Immune Defic Syndr. 2015;68:e39–41.
33. Oliviera M, Ibanescu RI, Anstett K, et al. Selective resistance profiles emerging in patients with cabotegravir, bictegravir, dolutegravir, and elvitegravir. Retrovirology. 2018;15:56.
34. McPherson TD, Sobleszczyk ME, Markowitz M. Cabotegravir in the treatment and prevention of human immunodeficiency virus-1. Expert Opin Investig Drugs. 2018;27:413–20.
35. Durham SH, Chahine EB. Cabotegravir-rilpivirine: the first complete long-acting injectable regimen for the treatment of HIV-1 infection. Ann Pharmacother. 2021;55:1–13.

36. Verloes R, Deleu S, Niemeijer N, Crauwels H, Meyvisch P, Williams P. Safety, tolerability and pharmacokinetics of rilpivirine following administration of the long-acting formulation in healthy volunteers. HIV Med. 2015;16:477–84.
37. Swindellis S, Andrade-Villanueva K, Richmond GJ, et al. Long-acting cabotegravir and rilpivirine for maintenance of HIV-suppression. N Engl J Med. 2020;382:1112–23.
38. Orkin C, Arasteh K, Gorgolas Hernandez-Mora M, et al. Long-acting cabotegravir and rilpivirine after oral induction for HIV-1 infection. N Engl J Med. 2020;382:1124–35.
39. Overton ET, Richmond G, Rizzardini G, et al. Long-acting cabotegravir and rilpivirine dosed every 2 months in adults with HIV infection [ATLAS-2M], 48 weeks results: a randomized, multicenter, open-label, phase 3b, noninferiority study. Lancet. 2021;396:1994–2005.
40. Jaeger H, Overton ET, Richmond G, et al. Long-acting cabotegravir and rilpivirine dosed every 2 months in adults with HIV-1 infection [ATLAS-2M], 96 week results: a randomized, multicenter, open-label, phase 3b, noninferiority study. Lancet HIV. 2021;8:e679–89.
41. Smith GHR, Henry WK, Podzamczer D, et al. Efficacy, safety, and durability of long-acting cabotegravir and rilpivirine in adults with human immunodeficiency virus type 1 infection: 5-yueras results from the LATTE-2 Study. Open Forum Infect Dis. 2021;8:ofab439. https://doi.org/10.1093/ofid/ofab439.
42. Howe ZW, Norman S, Lueken AF, et al. Therapeutic review of cabotegravir/rilpivirine long acting antiretroviral injectable and implementation considerations at an HIV specialty clinic. Pharmacotherapy. 2021;41:686–99.
43. Landovitz RJ, Donnell D, Clement ME, et al. Cabotegravir for HIV prevention in cisgender men and transgender women. N Engl J Med. 2021;385:595–608.
44. Delany-Moretlwe S, Hughes JP, Bock P, et al. Cabotegravir for prevention of HIV-1 in women: results from HPTN 084, a phase 3, randomized clinical trial, Lancet. 2022; [published online April1]. https://doi.org/10.1016/S0140-6736[22]00538-4
45. Iacob SA, Iacob DG. Ibalizumab targeting CD4 receptors, an emerging molecule in HIV therapy. Front Microbiol. 2017;8:2323.
46. Blair HA. Ibalizumab: a review in multidrug-resistant HIV-1 infection. Drugs. 2020;80:189–96.
47. Pace CS, Fordyce MW, Franco D, et al. Anti-CD4 monoclonal antibody ibalizumab exhibits breadth and potency against HIV-1, with natural resistance mediated by the loss of a V5 glycan in envelope. J Acquir Immune Defic Syndr. 2013;62:1–9.
48. US FDA. Trogarzo™ [ibalizumab-uiyk] injection, for intravenous use: US prescribing information. 2018. https://www.accseedata.fda.gov. Accessed 10 May 2022
49. Emu B, Fessel J, Schrader S, et al. Phase 3 study of ibalizumab for multidrug resistant HIV-1. N Engl J Med. 2018;379:645–54.
50. Meanwell NA, Krystal MR, Nowicka-Sans B, et al. Inhibitors of HIV-attachment: the discovery and development of temsavir and its prodrug fostemsavir. J Med Chem. 2018;61:62–80.
51. Li Z, Zhou N, Sun Y, et al. Activity of the HIV-1 attachment inhibitor BMS-626529, the active component of the prodrug BMS-663068, against CD4-independent viruses and HIV-1 envelopes resistant to other entry inhibitors. Antimicrob Agents Chemother. 2013;57:4172–80.
52. Nowicka-Sans B, Gong YF, McAuliffe B, et al. In vitro antiviral characteristics of the HIV-1 attachment inhibitor BMS-6265290, the active component of the prodrug BMS-663068. Antimicrob Agents Chemother. 2011;56:3498–507.
53. Lataillade M, Zhou N, Joshi SR, et al. Viral drug resistance through 48 weeks, in a phase 2b, randomized, controlled trial of the HIV-1 attachment inhibitor prodrug, fostemsavir. J Acquir Immune Defic Syndr. 2018;77:299–307.
54. Food and Drug Administration: RUKOBIA [fostemsavir] extended-release tablets for oral use. Available from: https://www.accessdata.fda.gov/drugsatfda docs/label/2020/212950s0001b1.pdf. Accessed on 11 May 2022
55. Muccini C, Canetti D, Castagna A, Spagnuolo V. Efficacy and safety profile of fostemsavir for the treatment of people with human immunodeficiency virus-1 [HIV-1]: current experience and place in therapy. Drug Des Devel Ther. 2022;16:297–304.

56. Lalezari JP, Latiff GH, Brinson C, et al. Safety and efficacy of the HIV-1 attachment inhibitor prodrug BMS-663068 in treatment-experienced individuals: 24 week results of Ai438011, a phase 2b, randomized controlled trial. Lancet HIV. 2015;2:e427–37.
57. Thompson M, Mendo F, Latiff G, et al. Long-term safety and efficacy of fostemsavir in treatment experienced HIV participants. Presented at the conference on retroviruses and opportunistic infections 2019, Seattle. March 4-7, 2019. Dent Abstr
58. Kozal M, Aberg J, Pialoux G, et al. Fostemsavir in adults with multidrug-resistant HIV-1 infection. N Engl J Med. 2020;382:1232–43.
59. Ackerman P, Thompson M, Molina J-M, et al. Long-term efficacy and safety of fostemsavir among subgroups of heavily treatment-experienced adults with HIV-1. AIDS. 2021;35:1061–72.
60. Seval N, Frank C, Kozasl M. Fostemsavir for the treatment of HIV. Expert Rev Anti-Infect Ther. 2021;19:961–6.
61. Molina JM, Segal-Maurer S, Stellbrink HJ, et al. Efficacy and safety of long-acting subcutaneous lenacapavir in phase 2/3 in heavily treatment-experienced people with HIV: week 26 results [Capella study]. J Int AIDS Soc. 2021;24(S4):75.
62. Molina JM, Yazdanpanah Y, Afani Saud A, et al. Islatravir in combination with doravirine for treatment-naïve adults with HIV-1 infection receiving initial treatment with islatravir, doravirine, and lamivudine: a phase 2b, randomized, double-blind, dose-ranging trial. Lancet HIV. 2021;6:e324–e33.

# Chapter 13
# New Antiviral Agents for Cytomegalovirus Diseases

## 13.1 Introduction

Cytomegalovirus (CMV) is a large DNA virus that commonly infects the world's population, 60% in developed nations and up to 90% in developing countries. Like all herpes group viruses, once infected it remains dormant in cells for life but can become reactivated with suppression of the immune system. Primary infection in normal hosts is usually asymptomatic or goes unrecognized, but occasionally presents with the mononuclear syndrome and mild hepatitis. However, vertical transmission from mother to child can result in congenital CMV with fetal abnormalities.

Severe CMV disease is commonly found in the severely immunosuppressed with T-cell disturbance and initially was commonly recognized during the early stages of the AIDS pandemic before widespread use of highly active antiretroviral therapy (ART), but now more commonly seen after stem cell and organ transplantations. The clinical spectrum of CMV disease in the immunosuppressed is wide with potential involvement of any organ: brain, bone marrow, eyes, gastrointestinal tract, liver, lungs, kidneys, heart, spinal cord, and peripheral nerves. Severe life-threatening diseases with high fatality without specific treatment are common in the immunocompromised hosts. Severe infection in these abnormal hosts results from primary infection transmitted by blood transfusion or organ donated and from reactivation in those latently infected.

Ganciclovir injections (approved in 1988) and oral valganciclovir (approved in 2001) have been the mainstay of treatment for CMV disease in the immunosuppressed patients for decades, with alternatives foscarnet (approved in 1991) and cidofovir (approved in 1993) reserved for ganciclovir failures. However, these drugs are limited by myelosuppression and nephrotoxicity. To overcome the toxic effects and emerging resistance to these older drugs, letermovir and maribavir were recently introduced to combat CMV infection in the abnormal hosts.



I. W. Fong, *New Antimicrobials: For the Present and the Future*, Emerging Infectious Diseases of the 21st Century,
https://doi.org/10.1007/978-3-031-26078-0_13

## 13.2 Letermovir

Letermovir (trade name Prevymis) is a newly approved antiviral for CMV infection. It was approved in the United States (US, in 2017), Canada, and the European Union for prophylaxis of CMV infection in hematopoietic stem cell transplant (HSCT). Letermovir (LTM) was derived from a new chemical class, the quinazolines, and targets CMV terminal complex protein pUL56 [1]. The viral terminal complex is highly specific to herpesviruses; it binds the newly synthesized CMV genome and the procapsid and uses ATPase to translocate the DNA into the capsid. LTM is highly specific for CMV and has no significant activity against other herpesvirus. It is more potent than ganciclovir against CMV, by 400-fold for the 50% effective concentration ($EC_{50}$) (4.5 nM versus 2 μM) and > 2000-fold the EC90 (6.1 nM versus 14.5 μM) [1, 2].

### 13.2.1 *Pharmacokinetics of Letermovir*

LTM is available as 240 mg and 480 mg for intravenous (IV) infusion and 240 mg and 480 mg tablets. The oral bioavailability in healthy subjects was 94%, but in HSCT recipients without cyclosporine it was 35%, and with cyclosporine it increases to 85% [3]. The median time for maximum serum concentration ($T_{max}$) after oral dosing ranges from 45 min to 2.25 h and taking LTM with food increases maximum serum concentration ($C_{max}$) by an average of 129.8%. The drug is highly protein bound (99%) with a mean terminal half-life of 12 h and is eliminated by hepatic uptake and excretion in the feces (93% with 73% unchanged drug) and urine excretion <2%, with minor oxidative metabolism [3].

No dose adjustment is recommended in patients with renal impairment with creatinine clearance (CrCl) >10 ml/min, and the safety in end-stage disease (CrCl <10 ml/min) is unknown. For the Prevymis injection, accumulation of the vehicle, hydroxyl betadex, could occur in patients with CrCl <50 ml/min. No dose adjustment is required in mild to moderate hepatic impairment but it is not recommended in severe impairment.

### 13.2.2 *Drug Interactions of Letermovir*

LTM is a moderate inhibitor of CYP3A, and drugs that are CYP3A substrates may have increased blood levels with co-administration with LTM, pimozide, and ergot alkaloids are contraindicated [3]. It is also a substrate of organic anion-transporting polypeptide, OATP1B1/3, and inhibitors may increase LTM plasma concentrations. LTM is also a substrate of CYP3A, CYP2D6, UGT1A1, and UGT1A3. Although oxidative metabolism is considered a minor elimination pathway, potent inducers such as rifampin are not recommended for co-administration. Co-administration of LTM with simvastatin or pitavastatin are not recommended and atorvastatin when cyclosporine is also being prescribed, due to increased levels of the statins and risk of myopathy and rhabdomyolysis. Co-administration of cyclosporine increases the blood levels of LTM. Several drugs can affect the blood levels of LTM, and it can also alter the blood levels of several drugs (for details see Ref. [3]).

### *13.2.3 Clinical Efficacy of Letermovir*

Prevention of CMV disease after HSCT or organ transplant has traditionally been by prophylaxis with ganciclovir or valganciclovir (with risk of myelosuppression) or preemptive treatment on detection of CMV DNAemia during regular monitoring. Preemptive monitoring and treatment has been favored after HSCT, while prophylaxis with valganciclovir for 3–12 months has been more commonly used after organ transplant.

Approval of LTM for CMV prophylaxis in HSCT was based on positive results of a phase III trial. In this double-blind, multicenter study, 565 patients (CMV seropositive) undergoing allogeneic HSCT were randomized to receive LTM or placebo orally or intravenously (IV) through week 14 after transplantation [4]. LTM was dosed at 480 mg daily or 240 mg daily in patients taking cyclosporine. Among 495 patients without CMV DNAemia at randomization by week 24, 122 of 325 (37.5%) patients receiving LTM developed clinically significant CMV infection versus 103 of 170 (60.6%) receiving placebo ($p < 0.001$). The frequency and severity of adverse events were similar between the 2 groups, including myelotoxic and nephrotoxic events. All-cause mortality at 48 weeks after transplantation was 20.9% in the LTM group and 25.5% in the placebo arm.

In a secondary analysis of patients with detectable CMV DNA ($n$ = 70), 48 received LTM and 22 placebo showed similar results as the primary analysis in participants with undetectable CMV DNA [5]. The incidence of clinically significant CMV infection was 26.1% lower with LTM at week 24, 64.6% with LTM versus 90.9% with placebo, $p = 0.01$.

Merck has also reviewed 48 observational studies that confirm the efficacy of LTM in preventing clinically significant CMV infection in allogeneic HSCT compared to controls (mostly preemptive therapy) up to 200 days in nearly 4000 transplant patients (presented at the European Society for Blood and Marrow Transplantation [EBMT] 48th Annual Meeting [Abstract #0S04-07], March 22, 2022).

In a retrospective cohort of hematopoietic cell transplant (HCT) with CMV seropositivity, LTM prophylaxis reduced the 1-year mortality by 79% compared to those not treated with LTM [6].

### 13.2.4 Safety of Letermovir

LTM has been generally well tolerated in the phase III and observational studies. Gastrointestinal side effects (nausea, vomiting, and diarrhea) were the most commonly reported, but others include fatigue, headaches, skin rash, and peripheral edema [7]. There is no nephrotoxicity or myelosuppressive effect with LTM, although a case of LTM induced hepatitis has been reported [7].

### 13.2.5 Letermovir Resistance

CMV resistance can be induced in vitro experimentally and has been reported in clinical settings. In experimental models, resistance to LTM can be induced after the third in vitro passage with escalating concentrations, compared to fifteen passages with foscarnet [8].

Resistance to LTM is due to mutations in the three genes that encode for viral terminase complex, most commonly UL51 (specifically at codon 231–369; e.g., V236M, L241P, R3695) and less commonly mutations at UL56 and UL89 [7]. LTM low barrier for resistance has been detected during clinical trials of LTM prophylaxis and has been reported just after 102 days of exposure during salvage therapy [9].

The degree of LTM resistance is not uniform and some mutations result in low-grade resistance and others with high-grade resistance, but high-grade resistance were more commonly associated with multiple mutations, i.e., UL51 P91s combined with UL56 S229F, L254F, and L2571 resulted in 290-fold increase in LTM resistance [7]. Moreover, LTM resistance would be less likely for prophylaxis where the viral replication and viral load are lower compared for treatment of immunosuppressed patients with CMV disease with higher viral replication and viral load.

### 13.2.6 Conclusion on Letermovir

LTM is a safe and effective prophylaxis for CMV infection in HSCT and is an alternative to preemptive therapy with CMV DNA weekly monitoring, but studies should be done to compare prophylaxis versus preemptive therapy with oral valgan ciclovir/ganciclovir or with LTM itself, including cost efficacy analysis. LTM is not approved for prophylaxis in solid organ transplant and a phase 3 study in renal transplant recipients is already underway.

Although LTM has been used off-label for treatment of CMV infections that failed standard therapy, the data is limited to few case reports with some success, but risk of resistance occurring [7]. A case series of 47 patients with established CMV infection treated with LTM from 13 transplant centers has been reported [10]. Most patients had intolerance to existing agents or resistance to treatment. Patients with

CMV viral load <1000 IU/ml were able to maintain suppression on LTM, but those with higher viral load had significantly lower success.

In vitro LTM in combination with standard DNA polymerase inhibitors (ganciclovir, foscarnet, and cidofovir) and the newly approved maribavir have shown synergy against CMV [7], but the need for combination therapy would need to be explored in severe or refractory CMV disease.

A major disadvantage of LTM is the low resistant barrier but prophylactic use may not lead to widespread resistance because of low viral load in these cases.

### 13.2.7 Maribavir for the Treatment of Cytomegalovirus Infection

Maribavir (marketed as Livtencity) is a novel agent that was approved by the FDA in December 2021 for the treatment of resistant/refractory CMV disease in post-transplant patients. It is a benzimidazole-1-riboside compound that competitively inhibits CMV UL97 protein kinase, essential for phosphorylating proteins needed for viral replication during the assembly within cells and release [11]. Since pUL97 is required for phosphorylation and activity of ganciclovir, co-administration of the two drugs result in antagonistic effect; however, synergy was observed with combination with cidofovir, foscarnet, and LTM [10].

Maribavir (MBV) has in vitro activity against CMV and Epstein-Barr virus (EBV) but not against herpes simplex virus 1 and 2 and varicella zoster virus (VZV); it inhibits CMV at <1 μM to ~15 μM [12].

### 13.2.8 Pharmacokinetics of Maribavir

MBV is available only in an oral formulation, and it is rapidly absorbed with $C_{max}$ in 1–3 h and bioavailability of 30–40%, but administration with high fat meal may decrease the plasma concentration by 30% [11]. Administration with antacids and low fat meal may not affect the drug exposure [13]. It is highly protein bound (97%), but can penetrate the blood-retinal barrier, and is extensively metabolized in the liver through the CYP3A4 pathway [12, 13]. It is primarily eliminated by biliary excretion, clearance not affected by renal impairment, and half-life is 5–7 h [13]. No dose adjustment is needed for severe renal impairment and moderate hepatic impairment.

MBV exposure is increased by 46% with co-administration of ketoconazole and decreased by 61% by rifampin; it increases tacrolimus exposure by 51% and does not affect the exposure of most other medications, including voriconazole [13].

Based on pharmacokinetic modeling, 400 mg twice daily would maintain free plasma concentration above the $EC_{50}$ of CMV for the entire dosing period, but for ≤200 mg dose the free concentrations were < 30 ng/ml which may be inadequate [14].

## *13.2.9 Clinical Efficacy of Maribavir*

### 13.2.9.1 Prophylactic Trials

Initial phase 2 studies with different doses of MBV, 100 mg twice daily, 400 mg daily, and 400 mg twice daily, found similar prevention of CMV infection (antigenemia or DNA PCR), but 400 mg twice daily was associated with the highest rates of dysgeusia and gastrointestinal side effects [12]. Thus, the phase 3 prophylactic study evaluated MBV at 100 mg twice daily. However, at this dose multiple randomized controlled trials of MBV failed to prevent CMV infection in HSCT and organ transplant recipients [12].

### 13.2.9.2 Preemptive Trials

A dose ranging phase 2, randomized, open-label study that included 82 HSCT and 77 organ transplant recipients with asymptomatic CMV viremia (1000–100,000 copies/ml) was conducted [15]. MBV at any dose (400–1200 mg twice daily) suppressed CMV viremia similar to valganciclovir, but two patients treated with MBV developed CMV recurrence with resistant mutations (UL97 T409M). Neutropenia occurred in 15–18% of valganciclovir-treated patients but only 4–5% of patients receiving MBV. However, the incidence of serious adverse events and discontinuation of trial medication was greater in the MBV group. A phase III randomized, double-blind, clinical trial is underway to compare MBV 400 mg twice daily versus valganciclovir for preemptive therapy of CMV infections in HSCT recipients.

### 13.2.9.3 Maribavir for Resistant or Refractory CMV Infections

MBV at different doses (400 mg, 800 mg and 1200 mg) twice daily was used in an open-label uncontrolled, randomized study in 120 HSCT or solid organ transplant recipients for refractory or resistant CMV infection [16]. There was no difference in response based on dose, and overall 67% had resolution of CMVemia at 6 weeks, but 35% developed CMV recurrence with 83% of these patients still receiving MBV. Mutations associated with MBV resistance (T4409M or H411Y) occurred in 13/25 (52%) of these patients.

In a phase 3, open-label study, 352 HSCT and solid organ transplant recipients were randomized to MBV 400 mg twice daily ($n$ = 235) or investigator-assigned therapy (IAT; ganciclovir/valganciclovir, foscarnet, or cidofovir) ($n$ = 117) for 8 weeks with 12 weeks follow-up [17]. The primary endpoint of CMV clearance at 8 weeks was achieved in 55.7% with MBV and 23.9% with IAT, $p < 0.001$; and secondary endpoint of maintenance of clearance and symptom control through week 16 was 18.7% vs 10.3% in favor of MBV. Adverse events were similar between the groups, but MBV was associated with less acute kidney injury than foscarnet (8.5% vs 21.3%) and neutropenia than ganciclovir/valganciclovir (9.4% vs 33.9%); less patients discontinued MBV (13.2%) than IAT (31.9%) due to adverse events. One per group died from treatment-related adverse events.

#### 13.2.9.4 Safety of Maribavir

Side effects with MBV were higher in phase 2 studies with higher doses, most commonly gastrointestinal in nature: dysgeusia (metallic or bitter taste) (65.0%), nausea (34.2%), vomiting (29.2%), diarrhea (23.3%), fatigue (20.8%), anemia (20.0%), peripheral edema (19.2%), headaches (1.58%), and renal impairment (15.8%) [12]. In phase 3 studies, similar side effects were reported but at lower rates with the current dosing; dysgeusia was the most frequent reported adverse event with MBV (8.5% vs 3.4%) which led to discontinuation in only 0.9% of patients [12].

### *13.2.10 Resistance to Maribavir*

Usually resistance to the agents that target CMV DNA polymerase (ganciclovir, cidofovir, and foscarnet) does not result in cross-resistance to MBV. However, cross-resistance between ganciclovir and MBV has been described with at least 10 mutations reported [18]. UL97 mutation at codon 342 (f342Y) led to ganciclovir resistance and low-level MBV resistance which led to a mutation at codon H411Y leading to MBV failure in 2 transplant recipients [19].

Resistance to MBV has been demonstrated in in vitro and in clinical trials caused by UL97 mutations, specifically mutations at codons 409, 411, and 480 are major causes of moderate-high grade MBV resistance after prolonged therapy [12]. In vitro, mutation of the UL27 gene can lead to low level MBV resistance [12].

### *13.2.11 Conclusion on Maribavir*

MBV is a safe oral alternative for refractory or ganciclovir-resistant CMV infections and offers a safer choice than foscarnet or cidofovir and may be used for multiresistant strains. It may also be effective for preemptive therapy in HSCT recipients, depending on the outcome of the ongoing trial. It is predictable that widespread use will lead to increased MBV-resistant CMV strains; thus the drug should be used judiciously.

# References


1. Ligat G, Cazal R, Hantz S, Alain S. The human cytomegalovirus terminase complex as an antiviral target: a close-up view. FEMS Microbiol Rev. 2018;42:137–45.
2. Lischka P, Hewlett G, Wunberg T, et al. In vitro and in vivo activities of a novel anticytomegalovirus compound A1C246. Antimicrob Agents Chemother. 2010;54:1290–7.
3. PREVYMIS™ [letermovir]. Reference ID 4179078. www.accessdata.fda.gov. Accessed 22 May 2022.
4. Marty FM, Ljungman P, Chemaly RF, et al. Letermovir prophylaxis for cytomegalovirus in hematopoietic-cell transplantation. N Engl J Med. 2017;377:2433–44.
5. Marty FM, Liungman P, Chemaly RF. et al, Outcomes of patients with detectable CMV DNA at randomization in the phase III trial of letermovir for prevention of CMV infection in allogeneic hematopoietic cell transplantation. Am J Transplant. 2020; https://doi.org/10.1111/ajt.15764.
6. Su Y, Stern A, Karatoni E, et al. Impact of letermovir primary cytomegalovirus prophylaxis on 1-year mortality after allogeneic hematopoietic cell transplantation: a retrospective cohort study. Clin Infect Dis. 2022;75:795–804.
7. Helou GE, Razonable RR. Letermovir for the prevention of cytomegalovirus infection and disease in transplant recipients: an evidence-based review. Infect Drug Resist. 2019;12:1481–91.
8. Chou S. Rapid in vitro evolution of human cytomegalovirus UL56 mutations that confer letermovir resistance. Antimicrob Agents Chemother. 2015;59:6588–93.
9. Turner N, Strand A, Grewal DS, et al. Use of letermovir as salvage therapy for drug-resistant cytomegalovirus retinitis. Antimicrob Agents Chemother. 2019;63:3.
10. Imlay HN, Kaul DR. Letermovir and maribavir for the treatment and prevention of cytomegalovirus infection in solid organ and stem cell transplant recipients. Clin Infect Dis. 2021;73:156–60.
11. Biron KK, Harvey RJ, Chamberlain SC, et al. Potent and selective inhibition of human cytomegalovirus replication by 1263W94, a benzimidazole L-riboside with a unique mode of action. Antimicrob Agents Chemother. 2002;46:2365–72.
12. Gandhi RG, Kotton CN. Evaluating the safety of maribavir for the treatment of cytomegalovirus. Ther Clin Risk Manag. 2022;18:223–32.
13. Song I, Ilic K, Sun K, Martin P. Clinical pharmacology of maribavir [SHP620]: a comprehensive overview. Biol Blood Marrow Transpl. 2019;25(Suppl):S342.
14. Marty FM, Boeckh M. Maribavir and human cytomegalovirus—what happened in the clinical trials and why might the drug have failed? Curr Opin Virol. 2011;1:555–62.
15. Maertens J, Cordonnier C, Jaksch P, et al. Maribavir for preemptive treatment of cytomegalovirus reactivation. N Engl J Med. 2019;381:1136–47.
16. Papanicolaou GA, Silveira FP, Langston AA, et al. Maribavir for refractory or resistant cytomegalovirus infections in hematopoietic-cell or solid organ transplant recipients: a randomized, dose-ranging, double-blind, phase 2 study. Clin Infect Dis. 2019;68:1255–64.
17. Avery RK, Alain S, Alexander BD, et al. Maribavir for refractory cytomegalovirus infections with or without resistance post-transplant: results from a phase 3 randomized clinical trial. Clin Infect Dis. 2021:ciab988. https://doi.org/10.1093/cid/ciab988.
18. Piret J, Boivin G. Clinical development of letermovir and maribavir: overview of human cytomegalovirus drug resistance. Antivir Res. 2019;163:91–105.
19. Strasfeld L, Lee I, Tatarowicz W, Villano S, Chou S. Virologic characterization of multidrug-resistant cytomegalovirus infection in 2 transplant recipients treated with maribavir. J Infect Dis. 2010;202:104–8.

# Chapter 14
# New Antiviral Agent for Influenza: Baloxavir

## 14.1 Introduction

Influenza seasonal outbreaks occur worldwide every year with estimates of 1 billion cases yearly with 3–five million severe cases which result in 290,000–650,000 deaths annually in the world [1]. The virus is a member of the orthomyxovirus family with four influenza types A to D, of which only A, B, and C cause human disease [2]. Influenza A and B are responsible for seasonal outbreaks and type A is responsible for pandemics, influenza C is rare and usually causes mild upper respiratory symptoms mimicking the "common cold."

The mainstay of prevention of seasonal influenza is by annual influenza vaccination, but the rate of vaccination in the general population is usually low (less than 40%) and the vaccine efficacy has been variable, depending on the match between the circulating virus and the strains used to make the vaccines. In a large study in the United States (US) over 3 seasons, the influenza vaccine efficacy in preventing serious influenza complications (i.e., pneumonia) requiring hospitalization was modest (38%) [3]. Thus, it is important to have an arsenal of effective antiviral agents for treatment and prevention of serious influenza infections in the at-risk individuals: the elderly, young children, pregnant women, patients with chronic comorbid illnesses, and the immunosuppressed.

## 14.2 Established Antivirals for Influenza Therapy

Amantadine was the first drug used for treatment and prevention of influenza A and was approved in the US in 1976 for influenza therapy but was previously approved by the Food and Drug Administration (FDA) in 1973 for the treatment of Parkinson's disease. Amantadine and rimantadine belong to the adamantanes class of drugs,

 
I. W. Fong, *New Antimicrobials: For the Present and the Future*, Emerging Infectious Diseases of the 21st Century,
https://doi.org/10.1007/978-3-031-26078-0_14

targeting the M2 ion channel protein of influenza A virus but not influenza B virus [4]. Amantadine and rimantadine are no longer recommended for prophylaxis or treatment of influenza A due to widespread resistance (>99%).

The neuraminidase inhibitors, oral oseltamivir and inhaled zanamivir, were approved in the US for influenza A and B infections in 1999, and intravenous peramivir was approved by the FDA in 2014. The neuraminidase inhibitors are currently first-line treatment for influenza in most countries of the world. They act by blocking the neuraminidase enzyme on the surface of influenza A and B, preventing release of new virus particles from the infected host (ciliated epithelia) cells, and resistance occurs much less readily than the adamantanes [5]. Prior to the influenza A (H1N1) pandemic in 2009, most circulating A strain were resistant to oseltamivir, but sensitive to zanamivir, but since then there has been very little resistance among recently circulating strains of influenza. Resistance is greater with oseltamivir (<3.5%) than zanamivir (<1%) and dual resistance is rare [6]. The majority of resistance is seen in influenza A (H1N1) in immunosuppressed patients treated with oseltamivir and less frequently with influenza H3N2 and B [5]. Resistance is rarely found in immunocompetent individuals.

Treatment of influenza with the neuraminidase inhibitors are most effective when started within 24 h of onset of illness, reducing the severity and duration by about 44% and 30% if given within 36 h of onset [5]. Oseltamivir was also shown to decrease the complications of influenza (otitis media, sinusitis, and pneumonia) [7]. As prophylaxis (shortly before or after exposure), oseltamivir has been shown to reduce the incidence of influenza by about 70–90% [5]. The neuraminidase inhibitors have been well tolerated with minor side effects.

## 14.3 Baloxavir a Novel Anti-Influenza Agent

Baloxavir (trade name Xofluza) was approved by the FDA for treatment (October 2019) and prophylaxis (November 2020) of influenza A and B infections. Baloxavir marboxil is a first-in-class prodrug that is metabolized by hydrolysis to the small active molecule (baloxavir acid) [8]. Baloxavir marboxil is a synthesized pyridine derivative of a polycyclic family [Wikipedia, Baloxavir marboxil].

The active molecule (baloxavir acid) inhibits the influenza cap-dependent endonuclease activity, used in "cap-snatching" by the virus polymerase complex essential in the life cycle, inhibiting viral mRNA synthesis and production of virions rapidly within 24 h [8]. As a consequence, a single dose is effective in blocking virus production and shorten symptoms. Baloxavir has antiviral activity against all four influenza subtypes including neuraminidase resistant strains with 90% effective concentration of 1.2–98.3 nmol/L [9].

## 14.4 Pharmacokinetics of Baloxavir

The pharmacokinetics of baloxavir varies with body weight (as body weight increases, the exposure decreases), but when dosed with the weight-based dosing, no significant difference in exposure was observed between different weight groups [10]. The recommended dose is 40 mg for subjects weighing less than 80 kg and 80 mg for those weighing 80 kg and more. Although baloxavir exposure is about 35% lower in non-Asians compared to Asians, this difference is not considered clinically significant.

The maximum serum concentration after dosing is achieved in 3.5–4 h ($T_{max}$) with mean peak concentration ($C_{max}$) of 96.4 ng/ml and area under curve ($AUC_{0-infin}$) of 6160 ng.hr./ml with 40 mg dose; and mean $C_{maxc}$ of 107 ng/ml and $AUC_{0-infin}$ of 8009 ng/ml with 80 mg dose [10, 11]. Administration with food decreases the $C_{max}$ by 48% and the $AUC_{0-infin}$ by 36%. A plasma concentration 24 h after dosing of ≥6.85 ng/ml was estimated to provide greater inhibition of influenza replication than oseltamivir; the 50% effective concentration of baloxavir acid for influenza A ranged from 0.63–0.77 ng/ml and for influenza B 2.7–4.1 ng/ml [12].

Baloxavir is highly protein bound, about 93%, with a mean terminal half-life of about 79 h (longer in Asians) and undergoes metabolism through UGT1A3 and CYP3A4 pathways with approximately 80% of the dose excreted in feces and 14.7% in the urine [10]. Although the effect of severe renal failure has not been studied, the pharmacokinetics should not be affected, and it is not affected by moderate hepatic failure, but it has not been evaluated in severe hepatic impairment.

## 14.5 Drug-Drug Interactions of Baloxavir

Baloxavir form a chelate with polyvalent cations, calcium, aluminum, magnesium, and iron, and co-administration result in decreased absorption [10]. It does not inhibit or induce cytochrome P450 enzymes or inhibit the UDP-UGT enzymes. Thus, no significant drug-drug interactions were noted with co-administration of baloxavir with itraconazole, probenecid, oseltamivir, midazolam, digoxin, and rosuvastatin [10].

## 14.6 Clinical Efficacy of Baloxavir

Baloxavir has been compared to placebo or neuraminidase inhibitor in three randomized, controlled trials (RCTs), and a systemic review and meta-analysis of these studies was recently published [13]. A total of 3771 patients were enrolled in the studies with baloxavir group, $n = 1451$; oseltamivir group, $n = 1288$; and placebo group, $n = 1032$. Baloxavir had a slightly shorter time than oseltamivir in relieving symptoms (mean − 1.29 h), but a significantly shorter time for alleviation of symptoms than placebo, −26.32 h. It was associated with a significant decline in influenza viral titers and viral RNA load than oseltamivir and placebo; and lower risk of adverse events than oseltamivir and placebo.

Another recent review and meta-analysis of antiviral agents for influenza, 26 trials with 11,897 participants, was published [14]. Of all treatments compared to placebo, zanamivir was associated with the shortest time for alleviation of symptoms, and baloxavir resulted in the lowest risk of influenza-related complications (risk ratio, 0.51; 95% CI, 0.32–0.80). Baloxavir was also associated with the lowest risk of adverse events, and oseltamivir (75 mg) was associated with the highest occurrence of nausea.

In a review of 32 studies, 7 studies on high-risk patients, the clinical efficacy of baloxavir was similar to the neuraminidase inhibitors but the mean decline in virus tiers after 24 h was significantly greater for baloxavir [15]. Drug-related adverse effects were also lower with baloxavir compared to oseltamivir. Treatment of an index case of influenza with baloxavir may result in greater reduction of intrafamilial transmission than oseltamivir [16], likely due its more rapid antiviral activity.

Compared to placebo and the neuraminidase inhibitors, baloxavir has been found to be very safe with less adverse events, possibly because of the single dosing.

## 14.7 Viral Resistance to Baloxavir

It is now evident that all classes of anti-influenza drugs have low genetic barriers to resistance: 1–2 amino acid substitutions can result in resistance; but other factors that may play a role is the strain of virus, resistance may emerge more frequently during treatment of A (H3N2) than A (H11N1); and low levels of circulating neutralizing antibodies may predispose to greater risk of drug resistance [17, 18]. Baloxavir resistance in previous phase 2 study was 2.2%, mainly due to infection with the 2009 pandemic influenza A (HiNi), but during the phase 3 trials (2014–2015), the antigenically drifted A (H3N2) emerged with baloxavir resistance of 7.9–9.7%. The amino acid substitution, PA-138 T, in the polymerase acidic protein was the most common mutation resulting in resistance and could be detected even 3 days after treatment and in most cases at 5 days [17]. Emergence of viruses with PA-138 T substitution was related to delay in symptoms alleviation and prolonged virus detection [18]. It is concerning that mutations with PA-138 T/M substitutions have been reported in 23% of children treated with baloxavir [19].

## 14.8 Role of Baloxavir in Influenza Treatment

It has been proposed that since baloxavir blocks virus replication rapidly and completely with a single dose, it may prevent transmission in the population. Using modeling, it has been estimated that accelerated baloxavir treatment of 30% of infected subjects within 48 h after symptoms onset, based on the 2017–2018 epidemic season, could have prevented 22 million infections and > 6000 deaths, with even greater impact with treatment within 24 h [20]. The major concern, however, is that mass treatment of the population with baloxavir could lead to increased resistant strains, making the drug ineffective to a large extent in a few years.

The Centers for Disease Control and Prevention (CDC) guidelines for antiviral treatment of influenza (2021–2022) recommend baloxavir only for suspected or confirmed uncomplicated influenza for outpatients in high-risk or healthy adults and

**Table 14.1** Indications for antiviral treatment and prophylaxis for influenza

| Treatment indications | Neuraminidase inhibitors | Baloxavir |
|---|---|---|
| Hospitalized patients | Oseltamivir or peramivir | No |
| Severe, complicated or progressive (outpatient) | Oseltamivir or zanamivir | No |
| High-risk patients: | Oseltamivir or zanamivir | Yes |
| – Chronic pulmonary disease | | |
| – Cardiovascular disease | | |
| – Malignancy | | |
| – Chronic renal insufficiency | | |
| – Liver disorders | | |
| – Diabetes and other metabolic disease | | |
| – Blood disorders (i.e., sickle cell disease) | | |
| – Immunosuppression (disease or medications) | | |
| – Neurologic and neurodevelopment disorders | | |
| – Subjects ≥65 years | | |
| – Residents of nursing home and chronic care facilities | | |
| – Pregnant women and within 4 weeks post-partum | | |
| – Children <5 years old | | |
| – Morbid obesity (body mass index ≥40) | | |
| – Indigenous peoples | | |
| Prophylaxis indications: | Oseltamivir or zanamivir | Yes |
| – High-risk patients | | |
| – Institution outbreaks (nursing homes, chronic care facilities, hospitals) | | |

Referenec [4]; Public Health Ontario; Antiviral medications for seasonal influenza: information for health care providers, 2019 children [4]. It was not recommended for hospitalized patients or outpatients with complications or progressive disease with suspected or confirmed influenza, and oseltamivir was recommended instead in these situations. This is due to the fact baloxavir has not been studied in patients with severe complicated influenza, and comparative studies versus oseltamivir are needed. Baloxavir would be a suitable agent for post-exposure prophylaxis in high-risk patients and may be more effective to prevent spread than oseltamivir in healthcare settings and outbreaks in nursing homes. Table 14.1 summarizes the indication of antiviral treatment for influenza and chemoprophylaxis.

New strategies for severe influenza are urgently needed as the neuraminidase inhibitors are only modestly effective. One randomized study compared the effect of neuraminidase inhibitors (primarily oseltamivir) in combination with baloxavir [21]. The combination therapy was not superior to single drug therapy but the combination was well tolerated. It would be interesting to determine whether the combination (even for 1 day) could decrease the risk of resistance to baloxavir, as this could be a strategy used for mass treatment in an influenza pandemic setting.

Baloxavir is not recommended in pregnant or nursing patients as there is lack of data on its safety in these settings and in the severely immunosuppressed subjects [4].

## References

1. Office WHOEMR. Global influenza strategy 2019–2030. Wkly Epidemiol Monit 2019.
2. World health Organization [WHO]. Influenza [Seasonal]. Bill World Health Organ; 2014.
3. Ghamande S, Shaver C, Murthy K, et al. Vaccine efficacy against acute respiratory illness hospitalizations for influenza-associated pneumonia during 2015–2016 to 2017–2018 seasons: US hospitalized adult influenza vaccine effectiveness network [HAVEN]. Clin Infect Dis. 2022;74:1329–37.
4. CDC. Influenza antiviral medications: summary for clinicians. Center for Disease Control and Prevention; 2021–2022.
5. Lampejo T. Influenza and antiviral resistance: an overview. J Clin Microbiol Infect Dis. 2020;39:1201–8.
6. Lee N, Hurt AC. Neuraminidase inhibitor resistance in influenza: a clinical perspective. Curr Opin Infect Dis. 2018;31:520–6.
7. Treanor JJ, Hayden FG, Vrooman PS, et al. Efficacy and safety of oral neuraminidase inhibitor oseltamivir in treating acute influenza: a randomized controlled trial. J Am Med Assoc. 2000;283:1016. https://doi.org/10.1001/jama.283.8.1016.
8. Noshi T, Kitano M, Taniguchi K, et al. In vitro characterization of baloxavir acid, a first-in-class cap-dependent endonuclease inhibitor of the virus polymerase PA subunit. Antivir Res. 2018;160:109–17.
9. Mishin VP, Patel MC, Chesnokov A, et al. Susceptibility of influenza A, B, C, and D viruses to baloxavir [1]. Emerg Infect Dis. 2019;25:1969–72.
10. Xofluza [baloxavir marboxil]—Accessdata. Fda.gov-US Food and Drug Administration; accessed June 21, 2022.
11. Kim Y, Lee S, Kim Y, Jang IJ, Lee S. Pharmacokinetics and safety of a novel influenza treatment [baloxavir marboxil] in Korean subjects compared to Japanese subjects. Clin Translat Sci. 2021;15:422–32.

12. Koshimichi H, Ishibashi T, Kawaguchi N, Sato C, Kawasaki A, Wajima T. Safety, tolerability, and pharmacokinetics of the new anti-influenza agent baloxavir marboxil in healthy adults: phase 1 study findings. Clin Drug Invest. 2018;38:1189–96.
13. Kuo Y-C, Lai C-C, Wang Y-H, Chen C-H, Wang C-Y. Clinical efficacy and safety of baloxavir marboxil in the treatment of influenza: a systematic review and meta-analysis of randomized controlled trials. J Micro Immun Infect. 2021;54:865–75.
14. Liu J-W, Lin S-H, Wang L-C, Chiu H-Y, Lee J-A. Comparison of antiviral agents for seasonal influenza outcomes in healthy adults and children. A systematic review and network meta-anlysis. JAMA Netw Open. 2021;4 [8]:e21119151. https://doi.org/10.1001/jamanetworkopen.2021.19151.
15. Taieb V, Ikeoka H, Wojciechowski P, et al. Efficacy and safety of baloxavir marboxil versus neuraminidase inhibitors in the treatment of influenza virus infection in high-risk and uncomplicated patients—a Bayesian network meta-analysis. Curr Med Res Opin. 2021;37:225–44.
16. Miyazawa S, Takazono T, Hosogaya N, et al. Comparison on intra-familial transmission of influenza virus from index patients treated with baloxavir marboxil or oseltamivir using an influenza transmission model and health insurance claims database. Clin Infect Dis. 2022;75:927–35.
17. Gubareva LV, Fry AM. Baloxavir and treatment-emergent resistance: public health insights and next steps. J Infect Dis. 2020;221:337–9.
18. Uehara THF, Haden FG, Kawaguchi K, et al. Treatment emergent influenza variant viruses with reduced susceptibility: impact on clinical and virologic outcomes in uncomplicated influenza. J Infect Dis. 2020;221:346–55.
19. Hirotsu N, Sakagushi H, Sato C, et al. Baloxavir marboxil in Japanese pediatric patients with influenza: safety and clinical and virologic outcomes. Clin Infect Dis. 2020;71:971–81.
20. Du Z, Nugent C, Galvani AP, Krug RM, Meyers A. Modeling mitigation of influenza epidemics by baloxavir. Nat Commun. 2020;!1:2750. https://doi.org/10.1038/s41467-020-165885-y.

# Chapter 15
# Direct Antiviral Agents for Hepatitis C

## 15.1 Introduction

The hepatitis C virus (HCV) was discovered in 1989 after intensive investigation in the 1970s to find the cause of post-transfusion non-A, non-B hepatitis [1]. Spontaneous clearance occurs in a minority of patients (20–30%); thus chronic HCV infection results in a large global burden and a major cause of liver cirrhosis, liver cancer, and need for liver transplantation. In 2015 the global burden of HCV was estimated to be 71 million people [2], but based on recent modeling, the global prevalence of HCV was estimated to be 0.7% or 56.8 million infections in 2020 [3], reduction largely due to the introduction of highly effective direct antiviral agents (DAA) in 2014. Regional prevalence rates were estimated to be greatest in Eastern Europe (2.9%) and Central Asia (2.6%) with the greatest numbers in South Asia (14.5 million) and East Asia (10.0 million). Egypt had the highest burden with prevalence of 14.7% in 2012, but with national programs for diagnosis and treatment, the prevalence of active infection (positive HCV-RNA) has declined to 9.5% [4].

There are seven major genotypes (GT) of HCV, with GT1 being the most common globally (46%), mainly in Europe, North America, and Australia, followed by GT3 (30%) mainly distributed in South Asia and GT4 prominent in the Middle East [5].

## 15.2 Treatment of Hepatitis C

The development of DAA for HCV is one of the greatest achievements in medical therapeutics, with the ability to cure >90–95% of chronic HCV with a combination of 2–3 agents given orally for 2–3 months. This is the first chronic viral infection that can be cured with medications. Prior to 2014, the treatment of choice was weekly

 
I. W. Fong, *New Antimicrobials: For the Present and the Future*, Emerging Infectious Diseases of the 21st Century,
https://doi.org/10.1007/978-3-031-26078-0_15

pegylated interferon alpha (PEG-IFNα) subcutaneously (sc) with oral ribavirin for 24–48 weeks with sustained long-term response of 40–80% and poorly tolerated [6]. Current DAA should be offered to all patients with hepatitis C with detectable virus (RNA) by PCR >6 months after initial infection. Treatment can clear the virus (cure) with improvement in liver histology, decrease fibrosis, occasionally reverse cirrhosis, decrease risk for hepatoma, and prolong life. Sofosbuvir, a nucleotide analogue and NS5B polymerase inhibitor, was the first DAA antiviral approved in the United States (US) in 2013 and Europe in 2014 for treatment of chronic HCV [1]. The development of fixed combinations of 2–3 drugs has resulted in improved cure rates and decreased drug resistance and enhanced the uptake of the DAA.

The DAA target various aspects of the HCV viral replication cycle, affecting key replication processes or structures by binding to components of the replicase complex or producing RNA chain termination; medications that target the NS3/4A protease contain the suffix "-previr," agents that inhibit the NS5B polymerase have suffix "-buvir," and NS5A inhibitors end in "-asvir" [7]. There were 13 DAA and 7 fixed combinations approved, but due to increased cure rates and stiff competition, 2 combinations and 5 DAA were voluntary withdrawn from the market. These include simeprevir, daclatasvir, Technivie (ombitasvir/paritaprevir/ritonavir), and Viekira Pak (ombitasvir/paritaprevir/ritonavir plus dasabuvir) (FDA-approved hepatitis C drugs, Verywell Health, updated January 13, 2020).

## 15.3 Combinations of Direct Antiviral Agents for Hepatitis C

### *15.3.1 Ledipasvir/Sofosbuvir*

Ledipasvir/sofosbuvir (marketed as Harvoni) was the first fixed combination approved in the US in 2014, used to treat genotypes 1, 4, 5, and 6. Sofosbuvir (SOF) is a NS5B polymerase inhibitor with high barrier of resistance and ledipasvir (LDV) is a NS5A protein inhibitor. Harvoni is administered as a single pill daily for mainly 8–12 weeks, consisting of SOF 400 mg with LDV 90 mg, and can be taken with or without food.

The maximum serum concentration ($C_{max}$) of LDV is achieved in 4–4.5 h after administration ($T_{max}$) and the $T_{max}$ of SOF was ~1 h after intake [8]. The plasma protein bindings of SOF and LDV are 61–65% and 99.8%, respectively. SOF is initially metabolized in the liver into the active nucleoside analogue, followed by dephosphorylation into the main inactive metabolite. SOF is mainly excreted in the urine (80%) and 14% recovered in the feces, with a median half-life ($t_{1/2}$) of 0.4 h [8]. LDV is excreted mainly in the feces with less than 14% in the urine with no significant metabolism and a median $t_{1/2}$ of 47 h [8].

LDV is an inhibitor of P-glycoprotein (P-gp) and breast cancer resistance protein (BCRP), while SOF is a substrate for P-gp and BCRP. LDV/SOF can be used in any degree of renal impairment (even on dialysis) without dose adjustment and in mild to severe hepatic impairment (Gilead Sciences, Harvoni drug information).

### 15.3.2 Efficacy and Indication of Ledipasvir/Sofosbuvir

HCV-GT1 can be treated with LDV/SOF for 12 weeks with or without compensated cirrhosis with sustained viral response at 12 weeks (SVR12) post-therapy of 97–99% [9]. Eight weeks of treatment can be used for non-HIV, non-Black patients without cirrhosis and with HCV RNA <six million IU/ml [9].

Twelve weeks of LDV/SOF has been shown to be effective with $SVR_{12}$ of 95% in 43 subjects with GT4 from 2 studies with or without compensated cirrhosis [9]. A larger study of 100 treatment-naïve Egyptian patients with GT4 showed 12 weeks LDV/SOF resulted in 99% SVR with minor side effects of 26% (headaches, fatigue, myalgia, and cough) [10].

LDV/SOF also has in vitro activity against HCV GT5 and GT6 and limited data on 12-week therapy of 41 GT5 and 25 GT6 patients with or without compensated cirrhosis resulted in 95–96% $SVR_{12}$ [9].

Several trials indicated that LDV/SOF with ribavirin (600 mg, increased as tolerated) for 12 weeks was an effective regimen for patients with decompensated cirrhosis and GT 1 or 4 [Patients with decompensated Cirrhosis. AASLD/IDSA; www.HCVGuidance.org on August 11, 2022].. The SVR12 ranged from 86% to 90%, which is lower than in patients with compensated cirrhosis treated without ribavirin.

### 15.3.3 Side Effects and Drug-Drug Interaction of Ledipasvir/Sofosbuvir

Overall Harvoni is well tolerated with minor side effects, cough, diarrhea, headache, irritability, fatigue, muscle ache, nausea, and insomnia. Reactivation of hepatitis B can occur in patients with coinfection. Drug-drug interactions may occur, and a list of drugs should be avoided: aluminum and magnesium containing drugs, some statins, amiodarone, digoxin, rifamycins, proton pump inhibitors, seizure medications, tenofovir disoproxil fumarate, warfarin, etc. (see list on Mayo Clinic drugs and supplements website).

### 15.3.4 Resistance to Ledipasvir/Sofosbuvir

Post-treatment resistance of HCV to LDV/SOF has been assessed in phase 2 and 3 clinical trials mainly for GT1. Virological failure was rare, 51 of 2144 (2.4%) treated patients and 74.5% with failure had detectable LDV-specific associated resistance mutations [11]. The percentage of virological failure and resistance progressively increased with longer duration of treatment. The common mutations were Q30R/H and/or Y93H/N in GT1a and Y93HH in GT1b; 35.3% of patients had two or more resistant substitutions with ≥100–1000 reduced susceptibility to LDV [11]. Only one patient in a phase 11 study with baseline LDV resistance developed resistance to SOF at failure, mutation at S282T.

### 15.3.5 Elbasvir/Grazoprevir

Elbasvir/grazoprevir (brand name Zepatier) is a fixed combination to treat HCV GT1 and GT4 with or without cirrhosis, approved in the US in January 2016. Elbasvir (EBR) is an NS5A inhibitor and grazoprevir (GZR) is a NS3/4A protease inhibitor combined in a single pill for once daily treatment with or without food for 12–16-week therapy [12]. Zepatier contains 50 mg EBR and 100 mg GZR and can be given with or without ribavirin, depending on genotype and previous treatment.

After administration, $T_{max}$ are reached at about 3 hours and 2 hours for EBR and GZR, respectively, and steady-state concentrations are achieved within 6 days [12]. EBR and GZR are highly protein bound, >99.9% and > 98.8%, respectively [8], but are distributed widely in most tissues (EBR) and mainly in the liver for GZR. Both agents are mainly metabolized by the cytochrome P450 system, with $t_{1/2}$ of 24 hours for EBR and 31 hours for GZR [12]. No dosage adjustments are needed for patients with severe renal impairment or mild hepatic dysfunction, but the drug should not be used in decompensated cirrhosis or moderate-severe hepatic impairment.

### 15.3.6 Drug Interactions of Elbasvir/Grazoprevir

EBR/GZR has many drug interactions: (i) potent P450 3A4 inducers (rifampin, phenytoin, carbamazepine, and St. John's wort) significantly reduce blood levels; (ii) strong P450 3A4 inhibitors (ketoconazole and ritonavir) may lead to increased concentrations, use with caution; (iii) transporting polypeptide 1 B 1/3 inhibitors (e.g., cyclosporine) increases GZR levels; (iv) nafcillin, bosentan, and modafinil may decrease blood levels; (v) HMG-CoA reductase inhibitors (statins) should be used at the lowest dose possible; (vi) HIV medications with atazanavir, darunavir, lopinavir, tipranavir, and efavirenz are contraindicated [12].

### 15.3.7 Clinical Efficacy of Elbasvir/Grazoprevir

A 12-week course of EBR/GZR has been evaluated in treatment-naïve patients with compensated cirrhosis or non-cirrhotic patients with GT 1a ($n$ = 211), GT 1b ($n$ = 171), GT4 ($n$ = 26), and GT 6 ($n$ = 13), with overall SVR12 of 95%, but only 80% SVR12 for GT 6 [13].

Combination of EBR/GZR with ribavirin has been used for experienced GT1 infected patients failing PEG-IFNα and ribavirin, 12-week therapy was effective in

92% (SVR12) even in patients with NS3 mutations, but only 82% when 14% possessed baseline NS5A-resistant substitutions, and triple therapy may be needed for 16 weeks [5]. Patients with GT4 who failed treatment with Peg-IFNα and ribavirin should also be treated with the combination with ribavirin for 16 weeks. EBR/GZR has also been found to be safe and highly effective in patients with end-stage renal failure on hemodialysis and HIV co-infected subjects.

### *15.3.8 Adverse Effects of Elbasvir/Grazoprevir*

EBR/GZR is generally safe and well tolerated; the most frequent side effects are headaches, nausea, fatigue, decreased appetite, anemia, fever, and elevations of liver enzymes [12]. Patients with coinfection can have reactivation of hepatitis B. Liver function tests should be monitored before and during treatment. The safety has not been established during pregnancy and breast feeding.

### *15.3.9 Resistance to Elbasvir/Grazoprevir*

Baseline resistance can affect the efficacy of EBR/GZR and are mainly associated with mutations to EBR, the NS5A inhibitor. In 617 HCV-GT1a-infected patients in Spain, resistance-associated substitutions to EBR was detected in 6.2%, most common were Y93C/H/N and Q30H/R [14].

Treatment emergent resistance with viral failures has been noted in phase 2 and 3 clinical trials. Resistant mutations were identified in 37 patients with GT1a, 8 patients with GT1b, and 5 patients with GT4 [12]. Treatment-emergent NS5A-related mutations (100% for GT1a and 1b) were more persistent than NS3-related substitutions (31% for GT1a and 50% for GT1b) at week 24 follow-up. NS5A mutations are more clinically relevant in leading to viral failure than NS3-related substitutions which may reduce the activity of GZR.

### *15.3.10 Sofosbuvir/Velpatasvir*

Sofosbuvir/velpatasvir (brand name Epclusa) was approved in the US in June 2016 for the treatment of the first six major HCV genotypes in adults, and in 2020 it was approved for children. The fixed combination of SOF (a NS5B polymerase inhibitor) with velpatasvir [VEL]), an NS5A replication complex inhibitor, is administered as one tablet daily for 12 weeks, with or without food, in patients with or without compensated cirrhosis.

Both SOF and VEL demonstrated antiviral activity against GT1–6 in HCV replicon assays and against resistance-associated variants related to other agents with different mechanisms of action, including NS3 protease inhibitors and NS5B non-nucleoside inhibitors [15].

The mean $T_{max}$ after oral administration for SOF and VEL were 0.5–1 hour and 3 hours post-dose, respectively [15]. VEL is more extensively bound to plasma protein than SOF, >99.5% versus 61–65%, and 98% of the dose present in plasma represent the parent compound. VEL undergoes slow metabolism by CYP enzymes (2B6, 2C8, 3A4), whereas SOF undergoes extensive first-pass hepatic metabolism. VEL is eliminated by biliary excretion with 94% excreted in the feces and SOF by renal elimination of the inactive metabolite. The $t_{1/2}$ of SOB is 0.5 hour and VEL 25 hours [8]. The fixed combination tablet once daily consist of 400 mg SOF and 100 mg VPR.

Dose adjustments are not needed for any degree of renal impairment and for mild to severe hepatic impairment.

### 15.3.11 Drug Interactions of Sofosbuvir/Velpatasvir

Concomitant use of drugs that are potent P-gp and CYP inducers are contraindicated (rifampin, rifabutin, carbamazepine, phenytoin, phenobarbital, St. John' wort), and moderate inducers are also not recommended (oxcarbazepine, modafinil, rifapentine, tipranavir/ritonavir, and efavirenz), as they decrease the plasma levels of SOF/VEL [15]. Gastric acid reducers (proton pump inhibitors, $H_2$-receptor antagonists and antacids) can reduce plasma levels of SOF/VEL and should be taken 4–6 hours apart and with food. Drug interactions may occur with co-administration of rosuvastatin, atorvastatin, digoxin, dabigatran etexilate, and tenofovir disoproxil fumarate.

### 15.3.12 Efficacy of Sofosbuvir/Velpatasvir

Randomized controlled trials have shown that SOF/VEL once daily for 12 weeks resulted in SVR12 of 95–99% in patients with GT1–6 infections with or without compensated cirrhosis [15]. Rate of response was similar in HIV co-infected patients. SOF/VEL plus weight-based ribavirin for 12 weeks provided SVR12 of 94% in patients with chronic HCV and decompensated cirrhosis. SVR12 may be slightly lower for GT3 with compensated cirrhosis (91%) and previously treated patients (89%). Patients who previously failed NS5A inhibitor-containing therapy may be treated with SOF/VEL plus ribavirin for 24 weeks [15].

### 15.3.13 Safety of Sofosbuvir/Velpatasvir

SOF/VEL has been well tolerated in clinical trials and the most common adverse events (not greater than placebo) were headache, fatigue, nasopharyngitis, and nausea [15]. Serious adverse events occurred from 1.5% to 5.5%, greater with combination with ribavirin and with 24 weeks of therapy. The most frequent serious adverse events were hepatic encephalopathy, sepsis, and anemia with ribavirin. In a large observational cohort from Taiwan, 3480 patients with HCV GT1–6 were treated with SOF/VPR for 12 weeks (±ribavirin) with an overall $SVR_{12}$ of 99.4% and adverse events of 10% and 0.6% were serious and one was related to treatment [16]. Patients co-infected with hepatitis B can have reactivation with increased liver enzymes.

### 15.3.14 Resistance to Sofosbuvir/Velpatasvir

In vitro, NS5B amino acid substitution at S282T was associated within 2–18-fold reduction in susceptibility to SOF in GT 1–6, and reduced VEL susceptibility was associated with NS5A substitutions at positions 24, 28, 30, 31, 32, 58, 92, and 93 in GT 1–6 [15]. In clinical trials the overall virological failure rate was low (20 of 1778 or 1.1%); single NS5A class resistance was observed at virological failure in 17 of the 20 patients [17].

### 15.3.15 Glecaprevir/Pibrentasvir

Glecaprevir/pibrentasvir (brand name Mavyret) was approved in the US in August 2017 for the treatment of chronic HCV GT1–6. Glecaprevir (GLE) is a potent HCV NS3/4 protease inhibitor and pibrentasvir (PIB) an HCV NS5A inhibitor, available in a fixed combination of 100 mg GLE and 40 mg of PIB in each tablet; usual treatment is 3 tablets daily (with food) for 8–16 weeks (Mavyret, NCBI Bookshelf, updated February 7, 2022). In vitro both agents have activity against the first 6 genotypes of HCV, but resistance appears rapidly with each individual agent, and this is prevented with the combination.

The $T_{max}$ of GLE/PIB is about 5 hours and food increases the absorption and both agents are highly protein bound (GLE 97.5%, PIB >99.9%) [8]. GLE is metabolized by CYP3A4 and PIB is not metabolized, both drugs weakly inhibit CYP3A4 and UGT. GLE is primarily excreted by the liver with 92.1% in the feces and $t_{1/2}$ of 6–9 hours; PIB is also primarily excreted in the feces (96.6%) with $t_{1/2}$ of 23–29 hours [8]. Both agents are inhibitors of P-gp, BCRP, and organic anion transporting polypeptide (OATP 1B/3 [18].

No dose adjustment is needed for patients with mild to severe renal impairment and mild hepatic impairment, but it is not recommended in moderate to severe hepatic impairment (Child-Pugh B & C). Data on Mavyret safety in pregnancy and lactating women are lacking.

### *15.3.16 Drug Interactions of Glecaprevir/Pibrentasvir*

Drugs that reduce GLE/PIB plasma concentrations and are not recommended for co-administration include rifampin, carbamazepine, efavirenz, and St. John's wort; agents that increase their concentrations and not recommended for co-administration are atazanavir, darunavir, lopinavir, and ritonavir [18]. Co-administration of Mavyret with statins increase their blood levels and risk of myopathy and should not be used together (atorvastatin, lovastatin, simvastatin) or use lower doses (rosuvastatin 10 mg maximum and 50% of pravastatin dose) [18]. It also increases plasma level of digoxin, 50% reduction in dose recommended, and dabigatran etexilate.

### *15.3.17 Efficacy of Glecaprevir/Pibrentasvir*

Phase II and III studies have shown that GLE/PIB is highly effective (≥95% SVR12) for patients with HCV GT1–6 for 8 weeks without cirrhosis and 12 weeks in those with compensated cirrhosis [18]. It is not indicated for patients with decompensated cirrhosis. Longer course of therapy (16 weeks) may be required for patients previously treated with an NS5A inhibitor [19].

A systematic review and meta-analysis of the real-world effectiveness of GLE/PIB in 12,531 adults with HCV reported SVR12 ≥ 95% across subgroups (HCV genotype, cirrhosis status, previous treatment, treatment duration, and subgroups of interest) [20]. Another recent review analyzed the efficacy in HCV patients previously treated with DAA [21]. Fourteen studies were included with 1294 patients mainly with GT 1–3, with one study with GT 4. Eleven of the studies used 12 weeks treatment and three used 12 or 16 weeks and only one report used rifabutin with GLE/PIB (clinical trial in the US). The overall SVR12 was 96% and in subgroup analysis longer treatment did not improve the outcome. Twelve weeks treatment was highly effective in patients with or without cirrhosis.

### *15.3.18 Adverse Effects of Glecaprevir/Pibrentasvir*

Mavyret has been well tolerated with the commonest side effect (greater than 10%) being headache and fatigue [18]. Similar to other DAA, it can precipitate reactivation of hepatitis B in those with combined infection. In the real-world experience, the most frequent side effects were pruritus, fatigue, and headaches, and serious adverse events were reported in 1% [20].

### *15.3.19 Resistance to Glecaprevir/Pibrentasvir*

In an analysis of resistance to GLE/PIB in 2200 patients treated for HCV GT 1–6, 22 patients experienced viral failure, and treatment-emergent mutations were detected in NS3 in 50% of patients and in NS5A in 82%, commonly as a combination of substitutions [22]. Treatment emergent NS3 substitutions were observed in 9 GT3a-infected patients (Y56H, Q80R, A156G, and Q168 [L/R]); 7 of the 17 GT3a-infected patients experiencing virologic failure had multiple substitutions in NS3. At the time of virological failure in 13 of 17 GT3a-infected patients, multiple substitutions in NS5A were found, most commonly linked substitutions of A30K plus Y93H detected in 10 patients.

### *15.3.20 Sofosbuvir/Velpatasvir/Voxilaprevir*

SOF/VEL/voxilaprevir (VOX) (brand name Vosevi) is a fixed triple combination in one pill approved by the FDA in July 2017 for chronic HCV GT1–6, previously treated with SOF but did not achieve clearance of the virus. This triple combination allows for inhibition of 3 non-structural proteins of HCV—SOF NS5A inhibitor, VEL NS5A inhibitor, and VOX NS3/4A inhibitor [Vosevi, NCBI Bookshelf, accessed July 29, 2022]. Each tablet contains 400 mg of SOF, 100 mg of VEL, and 100 mg of VOX, dosed at 1 tablet daily with food for 12 weeks.

The three agents in Vosevi have potent activity against the first 6 GT of HCV with mean 50% effective concentrations ($EC_{50}$) of SOF 14–110 nmol/L, VEL 0.33–6.6 nmol/L, and VOX 0.33–6.6 nmol/L [23]. After administration of the fixed combination tablet, the $C_{max}$ was reached in 2 hours for SOF and 4 hours for VEL and VOX with increased bioavailability for each component with food [23]. The protein binding for SOF is 61–65% and > 99% for VEL and VOX. Unlike SOF, VEL and VOX undergo slow metabolism by CYP enzymes, VEL by CYP2B6, CYP2C8, and CYP3A4 and VOX via CYP3A4 [23]. VEL and VOX are primarily excreted by the biliary tract and SOF by renal excretion with elimination half-life of about 17 hours, 33 hours, and 0.5 hour, respectively. VOX is a substrate and an inhibitor of the drug transporters P-gp, BRCP, OATp1B1, and OATP1B3.

SOF/VEL/VOX can be used in patients with any degree of renal impairment and mild hepatic impairment but should not be used in patients with moderate to severe hepatic impairment.

### 15.3.21 *Drug Interactions of Sofosbuvir/Velpatasvir/Voxilaprevir*

The drug interactions of Vosevi are similar to that of SOF/VEL outlined before [23, 24].

### 15.3.22 *Clinical Efficacy of Sofosbuvir/Velpatasvir/Voxilaprevir*

Randomized phase 111 clinical trials (POLARIS-1 and POLARIS-4) in patients with chronic HCV GT 1–6 previously treated with DAA, with or without compensated cirrhosis, showed SVR12 $\geq$ 95% [25]. In POLARIS-1, the most common NS5A inhibitors in the previously failed therapy were LDV (55%) and daclatasvir (23%), whereas in POLARIS-4 the majority of patients had failed a SOF containing regimen [25]. In DAA-naïve patients without compensated cirrhosis, SOF/VEL/VOX for 8 weeks was highly effective, but it was inferior to 12 weeks of SOF/VEL primarily due to the lower rate of SVR12 with GT1a.

The real-world experience with SOF/VEL/VOX as salvage therapy has been recently reviewed. In a meta-analysis of 15 studies with 1796 HCV-infected patients, the SVR12 was 93–96%, with significantly lower response in patients with cirrhosis and GT3 infected and previously treated with SOF/VEL [26].

### 15.3.23 *Safety of Sofosbuvir/Velpatasvir/Voxilaprevir*

The triple combination of SOF/VEL/VOX was generally well tolerated in the clinical trials. The most common side effects (>10%) were headache, fatigue, nausea, and diarrhea [25]. Serious adverse events occurred in $\leq$3%, none were treatment-related. In the real-world experience, the most frequent adverse events were similar and discontinuation due to side effects was very low, 0.66% [26].

### 15.3.24 *Resistance to Sofosbuvir/Velpatasvir/Voxilaprevir*

Although resistant mutants can be selected in cell culture to each component of Vosevi, emergent resistance-associated amino acid substitutions in patients with virological failure were rarely detected in clinical trials. In vitro, VOX had a higher resistant barrier than other HCV protease inhibitors [23]. Combination of multiple substitutions or mutations was more likely to result in failure. In 23 DAA-naïve patients with virological relapse, 1 patient with GT1a had emergent mutations, Q30R and L31M, in NS5A; among 7 DAA-experienced patients, 1 patient with

**Table 15.1** Features of the DAA combinations for hepatitis C

| Drugs | Brand Name | Indications | Duration | Cost/Mth US/Can. |
|---|---|---|---|---|
| LDV/SOF | Harvoni | GT1,4,5,6 | 8–12 wks | $19,019/22,333 |
| EBR/GZR | Zepatier | GT1 & 4 | 12–16 wks | $7675/18,674 |
| SOF/VEL | Epclusa | GT1-6 | 12–24 wks | $10,240/20,000 |
| GLE/PIB | Mavyet | GT1-6 | 8–12 wks | $20,124/20,000 |
| SOF/VEL/VOX | Vosevi | GT1-6 | 8–12 wks | $27,834/20,000 |

Abbreviations: *Mth* month, *EBR* elbasvir, *GLE* glecaprevir, *GT* genotype, *GZR* grazoprevir, *PIB* pibrentasvir, *SOF* sofosbuvir, *VEL* velpatasvir, *VOX* voxilaprevir
Price listings were obtained from the GoodRx for US prices and Ministry of Ontario drug benefits for the Canadian prices

GT4d had emergent substitution, Y93H in NS5A [25]. Pretreatment resistance-associated amino acid substitutions were not associated with virological failure, but GT3 and failed initial therapy with SOF/VEL were [26].

### *15.3.25 Summary*

Numerous trials have demonstrated that the 5 DAA combinations are highly effective and safe with mostly minor side effects. However, all the combinations have significant drug interactions with commonly used medications, and this should be verified before use. All of the combinations can be used in severe renal impairment but any protease inhibitor-containing regimens (e.g., glecaprevir, grazoprevir, and voxilaprevir) are not recommended for severe hepatic impairment (decompensated cirrhosis). Table 15.1 summarizes the features and cost of the DAA combinations.

## References


1. Manns MP, Maasoumy B. Breakthroughs in hepatitis C research: from discovery to cure. Nature Rev Gastroenterol Hepatol. 2022.; https://www.nature.com/articles/s41575-022-006068-8;19:533.
2. Polaris Observatory HCV Collaborators. Global prevalence and genotype distribution of hepatitis C virus infection in 2015: a modelling study. Lancet Gastroenterol Hepatol. 2017;2:161–76.
3. Polaris Observatory HCV Collaborators. Global change in hepatitis C virus prevalence and cascade of care between 2015 and 2020: a modelling study. Lancet Gastroenterol Hepatol 2022. https://doi.org/10.1016/S2468-1253[21]00472-6
4. El-Ghitany EM, Farghaly AG. Geospatial epidemiology of hepatitis C infection in Egypt 2017 by governate. Heliyon. 2019;5:e02249.
5. Tsukiyama-Kohara K, Kohara M. Hepatitis C virus quasispecies and genotypes. Int J Mol Sci. 2017;19:23. https://doi.org/10.3390/ijms19010023.
6. Fong IW. Major advances in hepatitis C treatment but not hepatitis B. In: Current trends and concerns in infectious diseases. Switzerland: Springer Nature; 2020. p. 103–29.
7. Kish T, Aziz A, Sorio M. Hepatitis C in a new era: a review of current therapies. Pharm Therapeut. 2017;42:316–29.

8. Smolders EJ, Jansen AME, ter Horst PGJ, Rockstroh J, Back DJ, Burger DM. Viral hepatitis C therapy: pharmacokinetic and pharmacodynamics considerations: a 2019 update. Clin Pharmacokin. 2019;58:1237–63.
9. AASLD-IDSA HCV Guidance Panel. Hepatitis C guidance 2018 update: AASLD-IDSA recommendations for testing, managing, and treating hepatitis C virus infection. Clin Infect Dis. 2018;67:1477–92.
10. Ahmed OA, Kaiser HH, Badawi R, et al. Efficacy and safety of sofosbuvir-ledipasvir for treatment of a cohort of Egyptian patients with chronic hepatitis genotype 4 infection. Infect Drug Resist. 2018;11:295–8.
11. Wyles D, Dvory-Sobol H, Svarovskaia ES, et al. Post-treatment resistance analysis of hepatitis C virus from phase 11 and 111 clinical trials of ledipasvir/sofosbuvir. J Hepatol. 2017;66:703–10.
12. Bell AM, Wagner JL, Barber KE, Stover KR. Elbasvir/grazoprevir: a review of the latest agent in the fight against hepatitis C. Int J Hepatol. 2016;2016:3852126. https://doi.org/10.1155/2016/3852126.
13. Zeuzem S, Ghalib R, Reddy KR, et al. Grazoprevir-elbasvir combination therapy for treatment naïve cirrhotic and noncirrhotic patients with chronic hepatitis C virus genotypes 1, 4, and 6 patients. Ann Intern Med. 2015;163:1–13.
14. Palladino C, Sanchez-Carrillo M, Mate-Cano I, et al. Low frequency of NS5A relevant resistance-associated substitutions to elbasvir among hepatitis C virus genotype 1a in Spain: a cross-sectional study. Sci Rep. 2017;7:2892. https://doi.org/10.1038/s41598-017-02968-7.
15. Craig SL. Sofosbuvir/vepatasvir: a review in chronic hepatitis C. Drugs. 2016;76:1567–78.
16. Cheng P-N, Mo L-R, TACR investigators. Sofosbuvir/velpatasvir for hepatitis C virus infection: real-world effectiveness and safety from a nationwide registry in Taiwan. Infect Dis Ther. 2022;11:485–500.
17. Hezode C, Reau N, Svarovskaia ES, et al. Resistance analysis in patients with genotype 1–6 HCV infection treated with sofosbuvir/velpatasvir in the phase III studies. J Hepatol. 2017;68:895. https://doi.org/10.1016/j.jhep.2017,11.032.
18. Abbvie. Mavyet [glecaprevir and pibrentasvir] Us Prescribing Information. Available from: https://www.accessdata.fda.gov/drugsatfda-docs/label/2019/209394s0081b1.pdf [2019]. Accessed August 2022.
19. Lok AS, Sulkowski MS, Kort JJ, et al. Efficacy of glecaprevir and pibrentasvir in patients with genotype 1 hepatitis C virus infection with treatment failure after NS5A inhibitor plus sofosbuvir therapy. Gastroenerology. 2019;157:1506–17.
20. Lampertico P, Carrion JA, Curry M, et al. Real-world effectiveness and safety of glecaprevir/pibrentasvir for the treatment of chronic HCV infection: a meta-analysis. J Hepatol. 2020;72:1112–21.
21. Shen C, Fan J, Dong C, et al. Efficacy and safety of glecaprevir/pibrentasvir in HCV patients with previous direct-acting antiviral therapy failures: a meta-analysis. Front Med 2020. https://doi.org/10.3389/fmed.2020.592472, https://doi.org/10.3389/fmed.2020.592472.
22. Krishnan P, Pilot-Matias T, Schnell G, et al. Pooled resistance analysis in patients with hepatitis C virus genotypes 1 to 6 infection treated with glecaprevir-pibrentasvir in phase 2 and 3 clinical trials. Antimicrob Agents Chemother. 2018;62:e01249–18.
23. Heo Y-A, Deeks ED. Sofosbuvir/velpatasvir/voxilaprevir: a review in chronic hepatitis C. Drugs. 2018;78:577–87.
24. Hong J, Wrigfht RC, Partovi N, Yoshida EM, Houssaini T. Review of clinically relevant drug interactions with next generation hepatitis C direct-acting antiviral agents. J Clin Translat Hepatol. 2020;8:322–35.
25. Xie J, Xu B, Wei L, Huang C, Liu W. Effectiveness and safety of sofosbuvir/velpatasvir/voxilaprevir as a hepatitis C virus infection salvage therapy in the real world: a systemic review and meta-analysis. Infect Dis Ther. 2022;11:1661–82.
26. Smith DA, Bradshaw D, Mbisa JL, et al. Real world SOF/VEL/VOX retreatment outcomes and viral resistance analysis for HCV patients with prior failure to DAA therapy. J Viral Hepatitis. 2021;28:1256–64.

# Chapter 16
# Antiviral Drugs for SARS-CoV-2 and COVID-19

## 16.1 Introduction

The emergence of severe acute respiratory syndrome coronavirus (SARS-CoV) in 2002 from the civet cat in China and Middle East respiratory syndrome coronavirus (MERS-CoV) in 2012 from camels in the Middle East [1] were presages of the emergence of severe acute respiratory syndrome coronavirus 2 (SARS-CoV-2) from the fish market of Wuhan, China, in December 2019. Since the initial epidemics of respiratory illness and pneumonia in China, SARS-CoV-2 have spread across the globe to produce the ongoing coronavirus disease 2019 (COVID-19) pandemic. This pandemic is still ongoing in its third year and will likely be entrenched as an endemic infection in the global communities like the influenza virus. As of March 2022, the World Health Organization (WHO) estimated that there were over 433 million confirmed cases and over 5.9 million deaths from COVID-19 [2], but many cases were never confirmed and the true global burden is more likely >600 million cases since the onset with >six million deaths.

SARS-CoV-2 is highly contagious and spread by close contact, respiratory droplets, and aerosols. Infection may be asymptomatic or produce mild respiratory illness in healthy adults and children but can produce severe pneumonia, sepsis-like syndrome, respiratory failure, and death in the vulnerable population: the elderly, obese, diabetics, patients with chronic pulmonary, kidney or cardiac diseases; the immunosuppressed, the poor, Black, Indigenous, and Latino people [3]. Even apparently healthy children can suffer severe consequences such as the multisystem inflammatory syndrome with myocarditis, and 30–60% of mild to severely ill patients may have persistent symptoms of fatigue, cough, and cognitive impairment for over 6–12 months (long COVID) [4].

The development of highly effective mRNA vaccines and others over a short period of time has been a remarkable achievement, but ongoing mutations of SARS-CoV-2 has rendered the vaccines relatively ineffective after 6 months of booster



I. W. Fong, *New Antimicrobials: For the Present and the Future*, Emerging Infectious Diseases of the 21st Century,
https://doi.org/10.1007/978-3-031-26078-0_16

doses [5]. Although vaccine pharmaceuticals are just about to release new vaccine boosters to cover the new omicron variants, new variants will appear with new mutations that will evade the antibodies. The monoclonal antibodies that were found effective earlier in the pandemic has been ineffective for the more recent omicron variants, and some were deauthorized by the FDA in January 2022. Thus, effective antiviral drugs are needed that are effective for various severity of illness caused by SARS-CoV-2. This chapter reviews the available antiviral agents on the market for treatment of COVID-19.

## 16.2 Coronaviruses

Coronaviruses are enveloped positive-sense RNA viruses, characterized by club-shaped spikes projecting from the surface to give the appearance of a solar corona, hence the name [5]. They are the largest group of viruses belonging to the *Nidovirales* order, which are characterized by large genomes for RNA viruses. The coronaviruses contain four main structural proteins: the spike, membrane, envelope, and nucleocapsid proteins. The trimeric spike glycoprotein is a class 1 fusion protein that mediates attachment to the host receptor [5]. The membrane protein is associated with assembly for viral particles; the envelope protein plays a role in pathogenesis by interacting with tight junction proteins; and the nucleocapsid protein plays a role in viral genome replication and cell signaling pathway [6].

The coronaviruses cause a wide variety of diseases in mammals, including humans, and birds and are classified into four genera based on phylogenetic analysis: alpha-coronaviruses, beta-coronaviruses, delta-coronaviruses, and gamma-coronaviruses [1]. Some coronaviruses, alpha- and beta-genera, cause human common cold symptoms and rarely severe respiratory disease. SARS-CoV, MERS-CoV, and SARS-CoV-2 belong to the beta-coronaviruses genera; SARS-CoV belong to lineage B and MERS-CoV to lineage C [1], and SARS-CoV-2 belong to a distinct lineage together with four horseshoe bat coronaviruses and a recently identified coronavirus from pangolins [7]. SARS-CoV-2 shares 79% genome sequences with SARS-CoV and 50% with MERS-CoV, but 96.2% with a bat coronavirus (RaTG13) detected in *Rhinolophus affinis* in Yunnan Province, China [7]. Bats are natural hosts of alpha-coronaviruses and beta-coronaviruses and believed to be the natural reservoir of SARS-CoV, MERS-CoV, and SARS-CoV-2, but the intermediate host that led to the COVID-19 pandemic by infecting humans remains unknown.

SARA-CoV-2 uses the same receptor as SARS-CoV for infecting the respiratory epithelium of humans, angiotensin-converting enzyme-2 (ACE2), and other animals (pig, rhesus monkey, ferret, civet, cat, dog, rabbit, and pangolin) [7]. However, MERS-CoV uses human and bat dipeptidyl peptidase-4 receptor for cell entry [1].

## 16.3 Antiviral Drugs for SARS-CoV

The first agents shown to be effective for hospitalized patients with severe COVID-19 infection were modifiers of the inflammatory response, dexamethasone [8], interleukin-6 inhibitors with conflicting data [9, 10], and Janus kinase 1 and 2 inhibitor [11]. Early administration of neutralizing monoclonal antibodies was also shown to decrease viral load and the frequency of hospitalization. Subsequently, specific antiviral agents were shown to be beneficial for moderate and severe COVID-19 infections.

## 16.4 Remdesivir

Remdesivir is a nucleotide prodrug (a phosphoramidate) that was originally developed to combat the Ebola outbreak in 2014 [12]. It has broad antiviral activity against RNA viruses, including many human and zoonotic coronaviruses. Remdesivir inhibits replication of RNA viruses by targeting the RNA-dependent RNA polymerase and is a delayed chain terminator [12]. Once inside cells, remdesivir is converted by intracellular kinases to its active nucleoside triphosphate metabolite (GS-441524) [13].

Remdesivir, given by intravenous (IV) administration, concentration in blood declines rapidly with plasma half-life ($T_{1/2}$) of ~1 hour; intracellularly it is rapidly converted to the active triphosphate analogue with a prolonged intracellular $T_{1/2}$ of 40 hours (peripheral blood mononuclear cell, PBMC) [13]. The maximum serum concentration ($C_{max}$) of remdesivir and GS-441524, 5.44 μg/ml and 0.15 μg/ml, respectively, exceed by many fold the concentration required to inhibit 90% of SARS-CoV-2 (IC90 1.76 μmol/L) [13]. The plasma protein binding of remdesivir is about 88% but that of the active metabolite (GS-441524) is low with >85% free fraction. Studies in monkeys demonstrate that active metabolite was detected in the testes, epididymis, eyes, and brain within 4 hours after dosing, lower in brain initially but accumulates over time [13].

The prodrug is a substrate for cytochrome P450 enzymes but rapidly metabolized by plasma hydrolases and exhibits low renal excretion (<10%), with 49% of the radio-labeled dose as GS41524 in urine [13]. The manufacturer does not recommend remdesivir use in severe hepatic and severe renal impairment, as no formal pharmacokinetic data are available. However, it has been used in several patients with acute or chronic severe kidney impairment, including patients on hemodialysis, with no significant toxic effects (Remdesivir: Drug information-UpTodate; August 30, 2022). It also has been used in pregnant and lactating women with SARS-CoV-2 infection and the pro-drug and the active metabolite can be detected in breast milk.

### 16.4.1 Animal Studies

Remdesivir has been evaluated in MERS-CoV- and SARS-CoV-2-infected rhesus monkeys. The dose of 5 mg/kg produced drug exposures equivalent to 100 mg used in humans. Compared to control monkeys, remdesivir-treated primates showed improved clinical and radiographic outcomes with undetectable virus in respiratory tract specimens 3-days post-inoculation [13].

### 16.4.2 Clinical Data on Remdesivir

Remdesivir, 200 mg IV loading dose then 100 mg daily for 10 days total, has been studied in several randomized, placebo, controlled trials (RCTs) with conflicting results in patients with moderate to severe COVID-19 infections admitted to hospital, requiring supplemental oxygen or mechanical ventilation. The National Institutes of Health (NIH)-sponsored phase 3 trial with 541 patients receiving remdesivir and 521 given placebo showed significant shortening of time to recovery, 10 days versus 15 days in favor of the study drug and a trend in improved survival at days 15 and 29 (11.4% vs 15.2%) [14]. The results of this trial led to the approval of remdesivir (Veklury) by the Food and Drug Administration (FDA) as the first antiviral treatment for COVID-19 in October 2020.

The WHO-sponsored RCT of hospitalized COVID-19 infected, 4 arm study, with 301 receiving remdesivir and 303 standard care, found no difference in outcome (need for ventilation and the time to discharge) or mortality [15]. Since then, there have been several RCTs with remdesivir in COVID-19-infected patients and two systematic reviews with somewhat different conclusions.

The first review included 9 studies with 1895 patients for qualitative synthesis: mean recovery time with remdesivir was 15.8 days and pooled mortality rate was 11.3% [16]. Treatment with remdesivir was associated with adverse effects warranting discontinuation of the drug in 17.8%. The meta-analysis of three RCTs indicated that remdesivir significantly reduced the mortality compared to placebo, $p \leq 0.001$.

In the most recent systematic review, data from five RCTs and two subtrials were reviewed, comparing 10-day treatment with remdesivir versus control standard care or placebo [17]. In hospitalized patients with COVID-19, the findings confirm that remdesivir resulted in little to no difference in mortality, may reduce the time to clinical improvement, and may lead to small reductions in serious adverse events. Although cost-effectiveness models assume remdesivir shortens duration of hospitalization, this may not be accurate in the real world. A large propensity-matched retrospective cohort from VA medical centers ($n$ = 2344) found remdesivir treatment was associated with prolonged hospitalization without improved survival [17]. It was also noted in one of the reviews that two randomized trials found no difference in outcome between 10 days or 5 days of remdesivir.

The benefit of remdesivir may be greater with earlier treatment before patients with COVID-19 require hospitalization. In a recent RCT, patients with COVID-19 symptom onset within 7 days and one or more risk factors for severe disease were randomized to IV remdesivir ($n$ = 279) for 3 days or placebo ($n$ = 283) [18]. COVID-19-related hospitalization occurred in 2 (0.7%) in the remdesivir group and in 15 (5.3%) of the placebo group, $p = 0.008$. The adverse events were similar between the groups and no deaths occurred. Thus, 3-day outpatient treatment with remdesivir was 87% effective in reducing hospitalization.

The recommended dosing for adults in hospitalized patients is 200 mg on day 1 followed by 100 mg daily for 4 days but may extend to 10 days with no substantial improvement by day 5 on mechanical ventilation. For nonhospitalized patients, 200 mg loading dose on day 1, then 100 mg on days 2 and 3.

### 16.4.3 Side Effects and Drug Interactions of Remdesivir

Remdesivir has generally been well tolerated. Adverse reactions >10% include increased serum glucose (3–11%) and increased creatinine (3–15%); those <10% include skin rash (<2%), nausea (3–7%), decreased hemoglobin (1–8%), lymphopenia (2%), prolonged prothrombin time (9%), increased liver enzymes (2–7%), and hypersensitivity reaction (<2%, including anaphylaxis); and rarely severe bradycardia, heart failure, hypotension, and acute hepatic failure (Remdesivir: Drug information—UpToDate, accessed August 30, 2022).

No formal studies have been performed on drug-drug interactions with remdesivir, and published guidance on drug interaction has been variable. Strong CYP3A34 inducers (rifampin, carbamazepine, phenytoin, etc.) may decrease the concentration of remdesivir and best to avoid. Chloroquine and hydroxychloroquine may also diminish the therapeutic effect of remdesivir and should also be avoided (Remdesivir: Drug information—UpToDate).

### 16.4.4 SARS-CoV-2 Resistance to Remdesivir

In vitro remdesivir drug-resistant viral populations can be selected by serial passaging of SARS-CoV-2 in the presence of remdesivir [19]. A single mutation in the RNA-dependent RNA polymerase (NSP12) at a residue conserved among all coronaviruses displayed decreased remdesivir sensitivity. However, there is no evidence of widespread transmission of remdesivir-resistant mutants of the circulating SARS-CoV-2 variants. A single case report of remdesivir-resistant mutation during therapy of a patient with acquired B-cell deficiency with persistent SARS-CoV-2 infection has been described recently [20]. A mutation, E802D, was identified in the nsp12 RNA-dependent RNA polymerase which conferred a ~ six-fold increase in remdesivir IC50.

### 16.4.5 Molnupiravir

Molnupiravir (brand name Lagevrio) was granted Emergency Use Authorization by the FDA in December 2021 for oral treatment of adults with mild to moderate COVID-19 with high risk of progression, within 5 days of symptom onset. It was approved in the United Kingdom a month before.

Molnupiravir is a small molecule ribonucleoside prodrug of N-hydroxycytidine (NHC) that inhibits the replication of RNA viruses, including influenza and the coronaviruses, with high resistant barrier [21]. After oral administration of molnupiravir, it is phosphorylated intracellularly to NHC triphosphate which is incorporated by the viral RNA by RNA polymerase leading to copying errors during viral replication, rendering the virus unable to replicate and noninfectious [22]. It has broad antiviral activity against several RNA viruses: SARS-CoV, MERS-CoV, SARS-CoV-2 (including remdesivir-resistant strains), bat-CoV (including SARS-like HKU3 and SHC014 and MERS-like HKU5), Venezuelan equine encephalitis virus, Eastern and Western equine encephalitis virus, chikungunya virus, and highly pathogenic influenza virus [23].

It was originally developed to treat influenza but was abandoned because of mutagenic effects; but based on available genotoxicity data and 5-day treatment, the FDA concluded that it has a low risk for genotoxicity [24].

After oral administration, the $C_{max}$ of molnupiravir is achieved between 0.25 and 0.75 h ($T_{max}$) and is converted by plasma esterase to the active antiviral NHC, which is widely distributed to body fluids and tissues and undergoes phosphorylation intracellularly to the triphosphate form by the host cell kinase [23]. The $T_{max}$ of NHC is 1–1.75 h with mean half-life of ~1 h, but with slower elimination following multiple or higher single doses (7.1 h at the highest dose tested) [25]. The amount of NHC excreted in the urine increases with the dose but maximally only 6.7%.

### 16.4.6 Animal Studies with Molnupiravir

The efficacy of molnupiravir has been studied in SARS-CoV-2-infected Syrian hamsters, ferrets, and mice implanted with human lung tissue models, as well as mice infected with MERS-CoV [23]. This agent was shown to decrease the viral load of the upper respiratory tract, block transmission to untreated contact animals, and prevent lung pathology of treated infected mice. Remdesivir failed to prevent SARS-CoV-2 shedding from the upper respiratory tract in rhesus macaque model [26].

### 16.4.7 Clinical Efficacy of Molnupiravir

Based on exposure-response analysis of phase 2 studies, 800 mg dose of molnupiravir (as four 200 mg capsules) was used for phase 3 trials. In a phase 3 double-blind, multicenter RCT, nonhospitalized, unvaccinated adults with mild to moderate confirmed COVID-19 within 5 days of onset and at least one risk factor for severe illness were randomized to molnupiravir 800 mg twice daily for 5 days ($n = 716$) or placebo ($n = 717$) twice daily for 5 days [27]. Molnupiravir significantly reduced hospitalization or death by day 29, 48 of 709 (6.8%) versus placebo, 68 of 999 (9.7%); with one death in the treated group and 9 in the control group. Adverse events were found in 30.4% of the molnupiravir group and 33.0% in the placebo group. Thus, molnupiravir was only 30% effective in reducing hospitalization or death in patients with COVID-19 not requiring hospitalization.

Data on the real-world efficacy of molnupiravir was recently provided from a study in Hong Kong during the omicron BA.2 wave [28]. This was a territory-wide retrospective cohort of patients hospitalized with confirmed SARS-CoV-2 infection without oxygen requirement. After propensity score matching, molnupiravir-treated patients ($n = 1856$) had all-cause lower mortality, 19.8 events per 10,000 person-days, than matched controls ($n = 1856$) with 38.06 events (50% reduction), $p = <0.0001$. Treated patients also had lower risk composite disease progression outcome and need for oxygen therapy.

### 16.4.8 Safety of Molnupiravir

Data from phase 1, 2, and 3 trials have found that molnupiravir was safe and well tolerated with no major safety concerns. The commonest side effects were nausea, diarrhea, and headaches [23], and allergic reactions can occur but are rare. The phase 3 study showed the adverse events in the molnupiravir group were similar to the placebo group. No formal drug-drug interactions have been performed but no significant drug interactions have been reported.

Molnupiravir is not recommended in pregnancy because fetal toxicity has been reported in animal studies; however, it may be considered in patients with high risk of progression and no alternative therapy are available, especially >10 weeks' gestation [29].

### 16.4.9 Resistance to Molnupiravir

SARS-CoV-2 genome is characterized by high error rates, short replication time, and abundant recombinations with swarms of variants with different degrees of fitness, which can rapidly develop drug resistance. However, molnupiravir appears to have a high genetic barrier to prevent viruses escaping from inhibition [30]. It was found to retain high antiviral activity against SARS-COV-2 with resistance to remdesivir and could inhibit variants in the Syrian hamster infection model [30]. In a recent in vitro study from Japan, the three antiviral agents maintain their activity with little change in $IC_{50}$ with omicron variants (BA.4 and BA.5) [31]. So far no resistance to molnupiravir has been reported in clinical studies or with real-world experience.

### *16.4.10 Nirmatrelvir/Ritonavir*

Nirmatrelvir/ritonavir (Paxlovid) was granted emergency use authorization by the FDA in adult and pediatric patients with mild-moderate COVID-19 infections, with high risk of progression for nonhospital use within 5 days of symptom onset in December 2021. It was also approved for use in many other countries.

Nirmatrelvir is a peptidomimetic, a covalent inhibitor, which binds and inhibits the main protease of SARS-CoV-2 to prevent replication [Wikipedia]. It has broad spectrum of activity against all known human coronaviruses, alpha- and beta-coronaviruses [32], including the new omicron variants [31].

Nirmatrelvir oral bioavailability is about 50%, and it is mainly metabolized by CYP34A; thus, it is co-administered with a low dose of ritonavir (a CYP3A4 inhibitor) to improve its pharmacokinetics [32]. Twice daily oral 300 mg nirmatrelvir with 100 mg ritonavir achieves and maintains plasma trough levels about 5–6 times the in vitro $IC_{90}$ of SARS-CoV-2 [33]. Nirmatrelvir undergoes minimal metabolism when administered with ritonavir with half-life of about 6 hours, eliminated mainly by the kidneys, protein binding of 69%, and $T_{max}$ 3 hours (Pfizer product monograph).

The usual dose in normal or mild renal impairment is nirmatrelvir 300 mg (2150 mg tablets) with ritonavir 100 mg twice daily for 5 days. For patients with moderate renal impairment, creatinine clearance ≥30 to <60 ml/min, the dose of nirmatrelvir is 150 mg with 100 mg ritonavir twice daily for 5 days [32]. Nirmatrelvir/ritonavir is not recommended for severe renal impairment (creatinine clearance <30 ml/min) or severe hepatic impairment. Although the manufacturer does not recommend it in pregnancy, it could be used if the benefits outweigh the risks as it is not absolutely contraindicated.

### *16.4.11 Drug-Drug Interactions of Nirmatrelvir/Ritonavir*

Ritonavir is a potent CYP3A inhibitor and may increase the plasma concentrations of drugs metabolized by CYP3A. Co-administration with Paxlovid is contraindicated for those drugs which elevated plasma concentrations may result in serious or life-threatening events (see product monograph for the list). Other CYP3A substrates may require dose reduction or monitoring of blood levels. Medications that inhibit or induce CYP3A may increase or decrease the concentration of nirmatrelvir/ritonavir, potentially leading to greater adverse effects or reducing the therapeutic effect. Ritonavir can inhibit CYP2D6 and induce other cytochrome enzymes and alter blood levels of drugs metabolized by these enzymes.

There is a large list of drugs which are contraindicated for use with nirmatrelvir/ritonavir including alfuzosin, ranolazine, antiarrhythmics (amiodarone, quinidine, etc.), fusidic acid, anticancer drugs (apalutamide, neratinib, venetoclax, etc.), anticonvulsants (phenytoin, carbamazepine, phenobarbital), anticoagulants (rivaroxaban), colchicine, rifampin, midazolam, statins (lovastatin and simvastatin), antipsychotics (lurasidone and pimozide), and vardenafil (Pfizer, Paxlovid product monograph).

### *16.4.12 Animal Models Treated with Nirmatrelvir/Ritonavir*

Mice infected with SARS-CoV-2 and treated with nirmatrelvir showed decreased pulmonary viral load, decreased inflammation, and lung injury [32].

### *16.4.13 Clinical Efficacy of Nirmatrelvir/Ritonavir*

A phase 2–3 double-blind RCT trial of 2246 symptomatic, unvaccinated, nonhospitalized adults with mild-moderate COVID-19 and at high risk of progression were randomized to nirmatrelvir/ritonavir ($n$ = 1120) or placebo ($n$ = 1126) for 5 days [33]. Treatment with nirmatrelvir/ritonavir was 89% effective in preventing progression to severe disease and reducing hospitalization and death compared to placebo. All 13 deaths occurred in the placebo group. The incidence of adverse events was similar between the two groups (22.6% for nirmatrelvir/ritonavir vs 23.95 for placebo). However, dysgeusia (5.6% vs 0.3%) and diarrhea (3.1% vs 1.65) occurred more frequently in the treated group.

The real-world efficacy of nirmatrelvir/ritonavir in hospitalized patients not requiring oxygen on admission with confirmed COVID-19 within 5 days of symptom onset was evaluated in a retrospective cohort from Hong Kong [28]. The study included 890 nirmatrelvir/ritonavir recipients and 890 matched controls. The antiviral treated group had lower risk of composite disease progression (HR 0.57) and need for oxygen therapy (HR 0.60) $p = 0.0001$. Also, the all-cause mortality rate was lower in nirmatrelvir/ritonavir patients, 10.8 events per 10,000 person-days compared to 26.47 events per 100,000 person-days in the control group, $p < 0.0001$. Time to achieving low viral burden was also significantly shorter in the treated group.

Data from Israel on the real-world experience with nirmatrelvir in an observational retrospective cohort during the omicron surge has just been published [34]. The study included 109,254 eligible patients with mild to moderate COVID-19 and high risk of progression, 3902 were treated with nirmatrelvir, and 105,352 were not treated. This population had high COVID vaccination rate with 1–2 booster doses. The study showed that nirmatrelvir significantly lowered the rate of hospitalization and death in patients 65 years of age and older (adjusted hazard ratio 0.27 and 0.21, respectively), but there was no evidence of benefit to younger patients.

### 16.4.14 Side Effects of Nirmatrelvir/Ritonavir

The major concern with the use of nirmatrelvir/ritonavir is the numerous drug-drug interactions. The controlled studies had reported that adverse events were no greater than the placebo groups. However, the common side effects noted were altered taste, diarrhea and muscle aches, loss of appetite, pruritus, abdominal pain, and occasionally liver disturbance (product monograph).

### 16.4.15 Rebound of COVID-19 after Antiviral Treatment

The Centers for Disease Control and Prevention (CDC) has recently released update to the public that 2–8 days after completing Paxlovid, a minority of patients may suffer from rebound of the virus with or without symptoms for a brief period. There is no evidence that this is related to drug resistance of the virus. Data from Pfizer indicate that rebound of the virus up to day 14 from onset of the phase 3 study was 2.3% in the nirmatrelvir/ritonavir group and 1.7% in the placebo group and was not associated with moderate-severe symptoms or resistance to nirmatrelvir [35].

A more formal study on the topic has been published as a preprint (before peer review), a retrospective cohort of US nationwide electronic health records of 13,644 adult patients treated with Paxlovid ($n$ = 11,270) or with molnupiravir ($n$ = 2374) [36]. The 7-day and 30-day rebound rates after Paxlovid treatment were 3.53% and 5.40% for COVID-19 infections, 2.31% and 5.87% for COVID-19 symptoms, and 0.44% and 0.77% for hospitalization. There were no significant differences in COVID-19 rebound after molnupiravir treatment: 5.86 and 8.59% for COVID infection, 3.75% and 8.21% for COVID-19 symptoms, and 0.84 and 1.39% for hospitalization.

### 16.4.16 Resistance to Nirmatrelvir

There is no clinical evidence of resistance to nirmatrelvir occurring after treatment but a distinct possibility exists. SARS-CoV-2 could acquire mutation in its main protease ($M^{pro}$) to develop resistance to nirmatrelvir, and using high-throughput protein design technique, a study estimated that ~40% of the designed mutations already exist in circulating lineages [37]. In vitro passaging of SARS-CoV-2 in increasing concentrations of nirmatrelvir results in multitude of $M^{pro}$ mutations via multiple pathways [38]. E166V mutation conferred the strongest resistance (~300-fold), but with loss of viral fitness. Most mutations confer low level drug resistance, but multiple accumulated mutations resulted in greater resistance.

## 16.5 Summary and Comments of Antiviral Agents for COVID-19

For patients with COVID-19 requiring oxygen in hospital, remdesivir is the only approved antiviral agent which shortens the course of illness but probably results in little or no difference in mortality. In patients with mild to moderate disease and risk factor for progression, early treatment within 5 days of symptom onset with all three of the antiviral agents is effective in reducing hospitalization and death. Nirmatrelvir and remdesivir are more effective than molnupiravir, which does not require IV therapy as remdesivir and lack the drug-drug interactions of nirmatrelvir/ritonavir. See Table 16.1 for summary of the three antiviral agents.

Based on retrospective data from Hong Kong, early treatment of hospitalized patients with COVID-19 (not initially requiring oxygen) with oral antivirals reduced mortality. It appears from the experience in Israel that nirmatrelvir has no significant benefit in young patients (under 65 years of age) fully vaccinated but with mild-moderate COVID-19 and high risk for progression. All three antivirals are well tolerated with mainly minor side effects; however, nirmatrelvir/ritonavir has many drug-drug interactions.

**Table 16.1** Summary of the features of the antiviral agents for COVID-19

| Features | Remdesivir | Molnupiravir | Nirmatrelvir/ Ritonavir |
|---|---|---|---|
| Type of agent | Nucleotide prodrug | Nucleoside prodrug | Peptidomimetic |
| Antiviral activity | Broad spectrum for RNA viruses | Broad spectrum for RNA viruses | Broad spectrum for RNA viruses |
| Mechanism | Chain terminator of RNA polymerase | Copying errors during viral replication | inhibit main protease |
| Drug interactions | Few | None | Many |
| Administration | Intravenous | Oral | Oral |
| Use in pregnancy | Yes | Contraindicated | If benefits outweigh risks |
| Indication | Hospitalized patients and high risk outpatients | High-risk outpatients | High-risk outpatients |
| Efficacy | Modest for inpatients 87% in preventing hospitalization | 30% in preventing hospitalization | 89% in preventing Hospitalization |
| Side effects | Minor, well tolerated | minor, well tolerated | minor, well tolerated |

Future studies in hospitalized patients with COVID-19 should compare the benefit of two antiviral agents (remdesivir and nirmatrelvir) versus remdesivir alone plus usual treatment with dexamethasone and anti-inflammatory agents.

## References

1. Fong IW. Emerging animal coronaviruses: first SARS and now MERS. In: Emerging Zoonoses. A worldwide perspective. Cham: Springer International; 2017. p. 63–80.
2. World Health Organization. COVID-19 weekly epidemiological update. 8 March 2022
3. Jost TS. Considering race and ethnicity in COVID risk assessments—legal concerns and possible solutions. N Engl J Med. 2022;387:481–3.
4. Huang L, Yao Q, Gu X, et al. 1-year outcomes in hospital survivors with COVID-19: a longitudinal cohort study. Lancet. 2021;398:747–58.
5. Sidik SM. Vaccines protect against infection from omicron subvariant—but not for long. Nature News; 18 March 2022
6. Triggle CR, Bansal D, Ding H, et al. A comprehensive review of viral characteristics, transmission, pathophysiology, immune response, and management of SARS-CoV-2 and COVID-19 as a basis for controlling the pandemic. Front Immunol. 2021;12:631139. 103389/fimmu.2021.631139
7. Hu B, Guo H, Zhou P, Shi Z-L. Characteristics of SARS-CoV-2 and COVID-19. Nature Rev Microbiol. 2021;19:141–54.
8. The Recovery Collaborative Group. Dexamethasone in hospitalized patients with COVID-19. N Engl J Med. 2021;384:693–704.
9. The TREMAP-CAP Investigators. Interleukin-6 antagonists in critically ill patients with COVID-19. N Engl J Med. 2021;384:1491–502.
10. Rubin EJ, Loggo DL, Baden LR. Interleukin-6 receptor inhibition in COVID-19---cooling the inflammatory soup. N Engl J Med. 2021;348:1564–5.
11. Kalil AC, Patterson TF, Mehta AK, et al. Baricitinib plus remdesivir for hospitalized adults with COVID-19. N Engl J Med. 2021;384:795–807.
12. Eastman RT, Roth JS, Brimacombs KR, Simeonov A, Shen M, Patnik S, Hall MD. Remdesivir: a review of its discovery and development leading to emergency use authorization for treatment of COVID-19. ACS Cent Sci. 2020;6:672–83.
13. Jorgensen SCJ, Kebriaei R, Dresser LD. Remdesivir: review of pharmacology, pre-clinical data, and emerging clinical experience for COVI-19. Pharmacotherapy. 2020;40:659–71.
14. Beigel JH, Tomashek KM, Dodd LE, et al. Remdesivir for the treatment of COVID-19—final report. N Engl J Med. 2020;383:1813–26.
15. WHO Solidarity Trial Consortium: Pan H, Peto R, Henao-Restrepo AM, et al. Repurposed antiviral drugs for COVID-19 –interim WHO solidarity trial results. N Engl J Med. 2021;384:497–511.
16. Bansal V, Mahapure KS, Bhurwal A, et al. Mortality benefit of remdesivir in COVID-19: systematic review and meta-analysis. Front Med 2021. https://www.frontiersin.org/articles/10.3389/fmed.2020.606429/full
17. Kaka AS, MacDonald R, Linskens EJ, et al. Remdesivir for adults with COVID-19: a living systemic review and meta-analysis for the American College of Physicians Practice points. Ann Intern Med 2022. https://doi.org/10.7326/M21-4784
18. Gottlieb RL, Veca CE, Paredes R, et al. Early remdesivir to prevent progression to severe COVID-19 in outpatients. N Engl J Med. 2021;386:305–15.
19. Szemiel AM, Merits A, Orton RJ, et al. In vitro selection of remdesivir resistance suggests evolutionary predictability of SARS-COV-2. PLOS Pathogens 2021. https://doi.org/10.1371/journal.ppat.1009929

20. GNDHI S, Klein J, Robertson A, et al. De novo emergence of a remdesivir resistant mutation during treatment of persistent SARS-CoV-2 infection in an immunocompromised patient: a case report. medRxiv. https://doi.org/10.1101/2021.11.0821266069
21. Agostini ML, Pruijssers AJ, Chappell JD, et al. Small- molecule antiviral β-d-N4-hydroxycytidine inhibits a proofreading-intact coronavirus with a high genetic barrier to resistance. J Virol. 2019;93:e01348–19.
22. Malone b, Campbell EA. Molnupiravir: coding for catastrophe. Nat Struct Mol Biol. 2021;28:706–8.
23. Khiali S, Khani E, Rouy SB, Entezari-Maleki T. Comprehensive review on molnupiravir in COVID-19: a novel promising antiviral to combat the pandemic. Future Microbiol. 2022;17:377–91.
24. Molnupiravir. COVID-19 treatment guidelines. NIH; Updated 24 Feb 2022. www.covid19treatmentguidelines.nih.gov.
25. Painter WP, Holman W, Bush JM, et al. Human safety, tolerability, and pharmacokinetics of molnupiravir, a novel broad spectrum antiviral agent with activity against SARS-CoV-2. Antimicrob Agents Chemother. 2021;66:e02428. 101128/AAC.02428-20
26. Williamson BN, Feldman F, Schwarz B, et al. Clinical benefit of remdesivir in rhesus macaques infected with SARS-CoV-2. Nature. 2020;585:273–6.
27. Bernal AJ, da Silva MGM, Musungaie DB, et al. Molnupiravir for oral treatment of COVID-19 in nonhospitalized patients. N Engl J Med. 2022;386:b509–20.
28. Wong CKH, Au ICH, Lau KTK, Lau EHY, Cowling BJ, Leung GM. Real-world effectiveness of early molnupiravir or nirmatrelvir-ritonavir in hospitalized patients with COVID-19 without supplemental oxygen requirement on admission during Hong Kong's omicron BA.2 wave: a retrospective cohort study. Lancet Infect Dis 2022. https://doi.org/10.1016/S1473-3099[22]00507-2
29. Fisher W, Eron JJ, Holman W, et al. Molnupiravir, an oral antiviral treatment for COVID-19. medRxiv. 2021; https://doi.org/10.1101/2021.06.17.21258639.
30. Tian L, Pang Z, Li M, et al. Molnupiravir and its antiviral activity against COVID-19. Front Immunol. 2022;13:855496. https://doi.org/10.3389/fmmu.2022.856496.
31. Takashita E, Yamayoshi S. Efficacy of antibodies and antiviral drugs against omicron BA.2.12.1, BA.4, and BA.5 subvariants. N Engl J Med. 2022;387:468–70.
32. Hung Y-P, Lee J-C, Chiu C-W, , Lee CV-C, Tsai P-J, Hsu I-L, Ko W-C. Oral nirmatrelvir/ritonavir therapy for COVID-19: the dawn in the dark.? Antibiotics [Basel] 2022; 11: 220
33. Hammond J, Leister-Tebbe H, Gardner A, et al. Oral nirmatrelvir for high risk, nonhospitalized adults with COVID-19. N Engl J Med. 2022;386:1397–408.
34. Arbel R, Sagy YW, Hoshen M, et al. Nirmatrelvir use and severe COVID-19 outcomes during the omicron surge. N Engl J Med. 2022;387:790–8.
35. Anderson AS, Caubel P, Rusnak JM, et al. For the EPIC-HR investigators. Nirmatrelvir-ritonavir viral load rebound in COVID-19. N Engl J Med. 2022;387:1047–9.
36. Wang L, Berger NA, Davis PM, Kelber DC, Valkow ND, Xu R. COVID-19 rebound after Paxlovid and molnupiravir during January–June 2022. medRxiv 2022. https://doi.org/10.1101/2022.06.21.22276724. [Preprint]
37. Padhi AK, Tripathi T. Hotspot residues and resistance mutations in the nirmatrelvir-binding site of SARS-CoV-2 main protease: design, identification, and correlation with globally circulating viral genomes. Biochem Biophys Res Commun. 2022;629:54–60.
38. Iketani S, Mohri H, Culbertson B, et al. Multiple pathways for SARS-CoV-2 resistance to nirmatrelvir. bioRxiv 2022. https://doi.org/10.1101/2022.08.07.499047. [Preprint]

# Part V
# Future Antimicrobials

# Chapter 17
# New Promising Antimicrobials in Development and Novel Approaches for Treatment of Infections

## 17.1 Introduction

Despite progress in developing and marketing new antimicrobials in the last several years, there is still an urgent need for developing new agents for the fight against multidrug-resistant (MDR) pathogens, especially MDR gram-negative bacilli (carbapenem-resistant *Acinetobacter baumannii*, *Pseudomonas aeruginosa*, and *Enterobacteriaceae*), MDR-*Neisseria gonorrhoeae,* MDR-*Mycobacterium tuberculosis* and nontuberculous mycobacteria, and others. It is of concern that of the 12 new antibiotics approved since 2017 only one was of a new class of drugs, lefamulin, a pleuromutilin.

In the past 2 decades, large corporations that were dominating antibiotic discovery and development have been leaving this area due to low profits and most companies working in this field are small biotechnical firms. It is worrisome that of the 15 new antibiotics approved by the Food and Drug Administration (FDA) in the past decade, 5 have been discontinued as the manufacturing companies have gone into bankruptcy or were sold off [1]. Antibiotics in general are not very expensive drugs, unlike other new medicines, and to make a profit by companies requires large sale volumes. But there is a paradox, for new antibiotics physicians in infectious diseases usually recommend limiting their use for specific indications, proven or suspected MDR-resistant bacteria which are relatively uncommon and, thus, result in low sales and low profits. To overcome this dilemma, public-private partnerships have been formed to provide funding for new antibiotic development. CARB-X (the Combating Antibiotic-Resistant Bacteria Biopharmaceutical Accelerator), founded in 2016, has provided $396 million in funding for 92 pre-clinical projects aimed at fighting drug-resistant infections (IDSA, Daily News Briefing, October 6, 2022). But will this be enough support for small biopharmaceutical companies? Only a few countries have implemented programs to provide support for new antibiotics development. The AMR Action Fund launched in June 2020 by the International



I. W. Fong, *New Antimicrobials: For the Present and the Future*, Emerging Infectious Diseases of the 21st Century,
https://doi.org/10.1007/978-3-031-26078-0_17

Federation of Pharmaceutical Manufacturers & Association and backed by large pharmaceutical companies will invest $1 billion over the next several years to help bring new antibiotics to the market (CDRAP News, Jan. 27, 2022).

The World Health Organization (WHO) has been leading the fight against MDR pathogens and reported that 163 countries have developed multisectorial AMR (antimicrobial resistance) national plans, but only 20% are actively monitoring their implementation.

## 17.2 New Antibiotics in Development

As of 2021, there were 77 new antibacterial agents in development, 45 traditional antibiotics and 32 are nontraditional [2]. Of the 45 traditional antibiotics, 27 (60%) are active against the WHO bacterial priority pathogens, 13 (28%) against *M. tuberculosis*, and 5 (11%) against *Clostridium difficile.* Of the 27 antibiotics against WHO priority pathogens, only two are active against MDR gram-negative pathogens and 40% are β-lactam-β-lactamase inhibitor combination with activity against metallo-β-lactamase producers.

Of the 32 nontraditional antibacterial agents, 6 are monoclonal antibodies, 9 are bacteriophages or phage-derived enzymes, 10 are microbiome modulating enzymes, 1 is an immunomodulating agent, and 6 are miscellaneous agents [2].

## 17.3 Traditional Antibiotics in Phase 3 Trials Against WHO Priority Pathogens

Sulopenem (intravenous [IV]/oral) is a synthetic penem, active against extended-spectrum β-lactamase (ESBL) producers, but not CRE (carbapenemase-resistant *Enterobacteriaceae*), and is being evaluated for complicated urinary tract infection (cUTi) and complicated intra-abdominal infections (cIAI), to provide early oral switch [2]. The FDA in July 2021 indicated that sulopenem requires further clinical trials to be undertaken [3].

Several new β-lactamase inhibitors (BLI) in combination with existing β-lactam agents are in development, most of them inhibit class A, C, and some D enzymes, but few inhibit class B enzymes [2]. Three of these combinations are in phase 3 trials durlobactam/sulbactam, taniborbactam/cefepime, and enmetazobactam [3]. Durlobactam is a modified diazabicyclooctane BLI with broader activity against class A, C, and D β-lactamases and appears to have activity against some *Enterobacterales* by binding to penicillin binding protein 2 (PBP2). It restores the activity of sulbactam against *A. baumannii* [2]. Two phase 3 trials in cUTI and one in HAP/VAP (hospital-acquired pneumonia/ventilator-associated pneumonia) are completed. The new combination was more favorable than colistin with respect to mortality, cure rate, and safety in the HAP/VAP trial [2]. Enmetazobactam is a derivative of tazobactam with enhanced bacterial cell penetration being studied in combination with cefepime in settings with high incidence of ESBL-producing *Enterobacterales* compared to piperacillin/tazobactam in cUTI. Results of this study (just published) showed the new combination was superior with respect to primary outcome of clinical cure and microbiological eradication [4]. Taniborbactam is a boronate-based BLI with activity against class A, B, C, and D β-lactamases, including metallo-β-lactamases (MBLs); in combination with cefepime, it is being studied in cUTi including acute pyelonephritis, compared to meropenem [2]. It is active against important carbapenem-resistant gram-negative bacilli, including *P. aeruginosa.*

Two agents developed to treat MDR-*N. gonorrhoeae* are in phase 3 trials, zoliflodacin and gepotidacin, the latter also being evaluated for UTI [3]. Zoliflodacin (oral agent) is a novel topoisomerase II inhibitor (spiropyrimidinetrione) with activity against *N. gonorrhoeae* and gram-positive cocci. It utilizes a distinct DNA gyrase binding site in GyrB compared to GyrA with the fluoroquinolones [2]. So far no cross-resistance with the fluoroquinolones has been detected. A multicenter trial of uncomplicated gonorrhea is ongoing to compare a single dose of zoliflodacin vs a single dose of ceftriaxone with azithromycin. Gepotidacin (IV/oral) is also a novel topoisomerase II inhibitor (triazaacenaphthylene) that inhibits bacterial DNA gyrase at a unique site of the GyrA subunit and the ParC subunit of the topoisomerase IV [2]. Some cross-resistance with the fluoroquinolones has been reported. Phase 3 trials are underway for uncomplicated gonorrhea and UTI. The prospect of this drug for future clinical marketing may be hampered by poor oral absorption and high incidence of diarrhea (95%) in phase 2 studies [2].

Two macrolide derivatives are undergoing development with activity against macrolide-resistant pneumococci and group A streptococci, solithromycin and nafithromycin. Solithromycin was being assessed in community-acquired pneumonia (CAP) and for treatment of gonorrhea, but the new drug application (NDA) was rejected by the FDA, because of inadequate data for liver toxicity [2]. Nafithromycin is undergoing a phase 3 trial for CAP in India.

An oral carbapenem, tebipenem, previously approved in Japan in 2009 for pediatric use, recently completed a phase 3 trial showing it was noninferior to IV ertapenem in cUTI [2]. It is active against ESBL-producing *Enterobacterales* but inactive against *A. baumannii* and *P. aeruginosa.*

### *17.3.1 New Antibiotic Being Developed in Phase 1 and 2 Trials*

There are several oral BLI and others for IV use in phase 1 trials, the most promising is QPX7728 (oral/IV), a boronate derivative which inhibits serine and MBLs of class A, B, C, and D in *A. baumannii, P. aeruginosa* and *Enterobacterales* [2].0.

A new semisynthetic tetracycline, KBP-7072 (oral), an aminomethylcycline for gram-positive respiratory pathogens and activity against *K. pneumoniae* and *E. coli* has completed 3 phase 1 trials [2]. It is effective against tetracycline-resistant pathogens with similar in vitro activity as tigecycline and omadacycline. Its potential benefit over the newer tetracyclines recently approved is unclear.

Novel classes of antibiotics are urgently needed rather than modifications of existing categories, which may facilitate greater ease for bacterial cross-resistance. Two new classes of antibiotics are in phase 1 or 2 studies, a Fabl (a critical enzyme for fatty acid biosynthesis in many bacteria) inhibitor, afabicin, and a FtsZ (filamenting temperature-sensitive Z, vital cell division protein conserved in most bacteria) inhibitor, TXA709 [2]. Afabicin (IV/oral), a pyrido-enamide, is a new *Staphylococcus*-specific antibiotic class with in vitro activity comparable to rifampin against intracellular and extracellular bacteria. It produces slow reduction of bacterial load and may be at risk for high level resistance. A phase 2 trial in skins and soft tissue infections due to *S. aureus* was completed, and there is plan for a phase 2 trial in bone and joint infections. The FtsZ inhibitor, TXA709 (IV/oral), is a methylbenzamide antibiotic targeting *S. aureus*, which is to be registered in phase 1 study.

Three polymyxin analogues (SPR-2006, QPX9003, and MRX-8) are in or completed phase 1 trials, being developed for MDR gram-negative infections, including *P. aeruginosa, A. baumannii,* etc., but their nephrotoxicity potential are to be determined [2].

Three antibiotic hybrids (antibiotic conjugated to functional moieties to create dual acting agents) or conjugates are in clinical development. The two phase 2 trials are TNP-2092 (IV/oral) and TPN-2198 (oral) [2]. TNP-2092 is a rifamycin-quinolizinone hybrid designed to reduce resistance to rifamycin, with comparable activity to rifampin being developed for staphylococcal prosthetic joint infections and gastrointestinal pathogens. TPN-2198 is a rifamycin-nitroimidazole hybrid with activity against anaerobes, *C. difficile* and *Helicobacter pylori*, and bacterial vaginosis is in a phase 1 trial in China.

There are five antibiotics being developed for treatment of *Clostridioides difficile* infections, four in phase 2 trials and one in phase 3 study. These are ridinilazole (oral), a bis-benzimidazole, in phase 3 trial; DNV-3837 (IV), an oxazolinone-quinolone hybrid, in phase 2 trial; MGB-BP-3, distamycin (oral, DNA minor groove binder), in phase 2 trial; ibezapolstat (oral), substituted guanine (DNA polymerase IIIC inhibitor) in phase 2 trial; and CR53123 (oral), diaryldiamine (methionyl-tRNA synthase inhibitor) in phase 2 trial [3].

### 17.3.2 *New Traditional Antibacterial Agents in Phase 1/2 Trials for* Mycobacteria Tuberculosis *and Nontuberculous Mycobacteria*

There are 14 antibacterial agents in clinical trials for treatment of mycobacteria infections, 13 for treatment of tuberculosis (TB) and 1 (SPR720) for lung infections caused by *Mycobacterium avium* complex or *Mycobacterium abscessus* [3]. Eight of the 14 antibacterial agents are from a new class and 9 have new antibacterial pharmacophores. Pharmacophore is part of the molecular structure responsible for the biological or pharmacological activity.

Examples of these new agents in phase 1–2 trials include TBAJ-587, a bedaquiline analogue with enhanced activity against *M. tuberculosis* and potentially lower cardiac side effects; GSK 2556286, proposed to act on cholesterol metabolism of the mycobacteria; and TBI-166 (pyrifazimine) now in phase 2 trial is a clofazimine analogue [3]. Two of these agents are diarylquinoline (similar to bedaquiline), three are oxazolidinone (similar to linezolid), and two are benzothiazinone (DprE1 inhibitor) [2], a new class of agents that appear in vitro to be more potent than isoniazid. Other novel class of agents include telacebec, an imidazopyridine amide that inhibits cytochrome bc1 in the respiratory cycle, and GSK3006656, an oxaborole that inhibits leucyl-tRNA synthetase [2].

## 17.4 Nontraditional Antibacterial Agents in Phase 3 Trials

There are six nontraditional antibacterial agents in phase 3 or later trials, three for *C. difficile*, and three *for S. aureus* infections [3]. The three agents for *C. difficile* infections are live biotherapeutic agents: BB128, a lyophilized donor fecal microbiota product to be given by colonoscopy; RBX2660, a liquefied donor fecal microbiota, to be given by enema; and SER-109, purified *Firmicutes* spores for oral administration, already completed one phase 3 trial. Thus, two of the three biotherapeutic agents are not really novel agents.

The three nontraditional agents for treatment of *S. aureus* infections are tosatoxumab, an immunoglobulin M (IgM) for IV therapy; exebacase, a phage-derived endolysin recombinant protein for IV; and reltecimod (IV), an immune modulator (CD28 T-lymphocyte receptor mimetic) [3].

### *17.4.1 Nontraditional Antibacterial Agents in Phase 1 and 2*

There are 25 nontraditional antibacterial agents in phase 1 and 2 trials [3] and 31 overall in clinical development [3]. Five of these are bacteriophages or their products (endolysin), three are for inhalation and two for IV therapy targeting *S. aureus, E. coli*, and *P. aeruginosa*. Eight of the agents are for *C. difficile* infection (all oral), four live microbes, two antibiotic inactivator (one combined with protective colon-target adsorbent), one polyclonal antibody, and one synthetic glycan.

Three monoclonal antibodies are being studied, two for *S. aureus* (1 combined with rifamycin) and one for both gram-positive and gram-negative bacteria. Novel agents being developed include a broad-spectrum anti-toxin agent and nanoparticle (IV) for *P. aeruginosa, A. baumannii, Enterobacterales, S. aureus*, and *S. pneumonia* infections; amido piperidine (inactivation of TetR-like repressor, EthR2) for oral treatment of TB; and a recombinant human plasma gelsolin protein (Rhu-pGSn) (IV) for nonspecific gram-positive and gram-negative bacterial infections [3].

## 17.5 New Antifungal Agents in Development

The emergence of antifungal resistance and new fungal pathogens needs global attention and development of new agents to address this issue. New antifungals approved in the last several years have been modifications of existing classes with no major advantage over older agents. Thus, novel agents are needed with different modes of action to strengthen our antifungal armamentarium. The increasing emergence of azole- and echinocandin-resistant yeasts (*Candida glabrata, Candida auris,* etc.) and azole-resistant *Aspergillus* spp. are the most evident, but there are obscure mold infections, being increasingly recognized in the immunosuppressed, where there are no active reliable agents (*Fusarium* species and *Lomentospora* species) [5].

However, there is hope as new antifungal agents being developed appear to be very promising. Seven new agents in the antifungal pipeline has recently been reviewed [6]. Although three of these are modification of existing classes, four represent novel classes of antifungal agents. Currently, we rely on four classes of antifungals in our arsenal: the polyenes (amphotericin B, nystatin), azoles (fluconazole, voriconazole, itraconazole, posaconazole, and isavuconazole), echinocandins (anidulafungin, caspofungin, and micafungin), and the pyrimidine analogue 5-flucytosine.

### *17.5.1 Derivatives of Existing Class of Antifungals*

Oteseconazole (oral), a new tetrazole with greater affinity for the target enzyme (14-alpha demethylase) than the triazoles (i.e., fluconazole), was designed for greater selectivity, fewer side effects, and enhanced efficacy compared to present azoles [6]. It may have less drug-drug interactions than other members of this class, as the affinity for fungal CYP51 is >2000-fold less than for human cytochrome enzymes. It has in vitro activity against *Candida* species (including fluconazole-resistant *Candida krusei* and *C. glabrata)*, dermatophytes, *Coccidioides*, and selected *Mucorales* species. It is currently in phase 3 study for recurrent vulvovaginal candidiasis (VVC) compared to fluconazole. Its advantages over current azoles include better activity than fluconazole, could potentially be used for candidemia/invasive candidiasis, and safer than voriconazole/ posaconazole.

Rezaconazole (IV) is a new azole with activity against fluconazole- and echinocandin-resistant *Candida* species (including *C. auris)*, *Aspergillus*, *Pneumocystis jiroveci*, and *Cryptococcus* [6]. Its unique pharmacologic feature allows for once weekly dosing, and it is in phase 3 trial for invasive candidiasis.

An encochleated amphotericin B (Matinas), C-AmB, in phase 1 and 2 studies, was designed as a novel lipid nanocrystal for oral therapy of serious fungal infections [6]. The in vitro activity of C-AmB is similar to the parent compound, and it is expected to be safer than amphotericin B deoxycholate (AmB). In phase 1 studies, C-AmB was well tolerated, and the usual side effects of the parent compound was not noted (hypokalemia, anemia, and renal dysfunction). Phase 2 studies in immunosuppressed subjects with refractory candida esophagitis showed dramatic results. A phase 2 trial in acquired immunodeficiency syndrome (AIDS)-associated cryptococcal meningitis is underway to compare oral C-AmB versus parenteral AmB for induction therapy.

### 17.5.2 Novel Antifungal Agents

Fosmanogepix (Amplyx) is the precursor of the active compound manogepix, a first-in-class antifungal that blocks GP1 (glycosylphosphatidylinositol) production via inhibition of Gwt1, important for cell wall construction and maintenance [6]. It has broad antifungal activity against yeasts and molds, with *Candida* minimum inhibitory concentrations (MICs) much lower than current azoles and echinocandins, including *C. auris,* but elevated against *C. krusei.* It is active against *Cryptococcus neoformans,* non-*Candida* yeasts (*Malassezia* and *Trichosporon*), unlike the echinocandins, *Aspergillus* species, *Fusarium, Scedosporium,* and *Lomentospora,* but limited activity against *Mucorales* spp. Animal models of disseminated candidiasis with azole-resistant strains have demonstrated good efficacy with good penetration in the eyes and meninges (unlike echinocandins). In mouse models of cryptococcal meningitis, fosmanogepix was comparable to fluconazole, but in the mouse model of pulmonary *Coccidioides*, it showed better survival than fluconazole. In immunosuppressed animal models of pulmonary infection with *Scedosporium, Lomentospora*, and *Rhizopus*, it also showed good activity. Phase 1 study showed that fosmanogepix oral doses were > 90% bioavailable. A phase 2 study in candidemia (single arm) of 20 patients showed 80% treatment success; no significant adverse effects were noted [6].

Olorofim is a first in class novel antifungals of orotomide, a pyrimidine synthesis inhibitor [6]. It has high potency against *Aspergillus* species, greater than current agents including azole-resistant strains, and rare molds (*Talaromyces, Trichophyton, Alternaria, Fusarium, and Penicillium*) infections. It showed activity against *Scedosporium* and *Lomentospora* species, frequently resistant to all other antifungals. Its anti-mold activity is greater than the triazoles and AmB, but it lacks activity against yeasts and *Mucorales*. Animal models of invasive aspergillosis showed similar response as posaconazole, and in meningeal infection with *Coccidioides,* it showed persistent suppressed fungal burden longer than seen with other agents [6]. Phase 1 studies of IV and oral formulations showed good tolerance, and a phase 2b open-label study is ongoing for rare fungal pathogens lacking other treatment options. Successful treatment has been reported with two cases of *Lomentospora* infections and refractory disseminated coccidioidomycosis failing multiple antifungal therapies, in combination with posaconazole [6].

Ibrexafungerp is a first in class of the triterpenoid antifungals or oral glucan synthase inhibitor (same target as the echinocandins), a semisynthetic derivative of enfumafungin [6]. Its mechanism of action and structure are similar to the echinocandins, and it is developed for IV and oral administration. The activity in vitro includes azole-resistant strain of *Candida* species (*C. glabrata* and *C. auris*) and *Aspergillus* and *P. jiroveci*, but it lacks activity against *Fusarium* and *Mucorales*. It has completed several phase 2 and 3 trials in VVC (superior to fluconazole with lower recurrence at 4 months). A phase 2 study has been done for invasive candidiasis, as step-down therapy after initial echinocandin and a large randomized trial is being planned for this condition comparing ibrexafungerp to fluconazole as step-down therapy. An open-label study for the treatment of *C. auris* is near completion [6]. The side effects reported in these studies have been mainly limited to the gastrointestinal tract, nausea, vomiting, abdominal bloating, and diarrhea.

ATI-2307 is novel arylamidine compound with new mechanism of action among antifungal agents, by inhibiting mitochondrial function, causing collapse of membrane potential in fungal cell preferentially [6]. It remains in tissues for weeks, and there is little metabolism of the drug; it is only available for IV administration and has low potential for drug interaction. ATI-2307 has broad antifungal activity against *Candida* species (including fluconazole- and echinocandin-resistant strains and *C. aurris*), *Cryptococcus, Aspergillus,* and *Fusarium.* In animal models of systemic candidiasis, it demonstrated lower effective minimum dose than fluconazole, micafungin, and AmB. Phase 1 studies showed no serious side effects and these include neurosensory symptoms, tachycardia during infusion, headache, and dysgeusia. A phase 2 study is being planned for to compare an echinocandin to ATI-2307 in treatment of candidemia with antifungal resistance and due to *C. auris*.

## 17.6 New Anti-Parasitic Agents in Development

Traditionally, there have been little incentives by pharmaceutical companies to develop new drugs for parasitic diseases, especially those considered neglected tropical diseases. The exception in recent years has been the development of new antimalarial agents. However, researchers in Chagas disease are excited with the discovery of a novel agent for this disease which is a major cause of chronic cardiomyopathy in Latin America.

The benzoxaborate prodrug AN15368 is activated by parasite carboxypeptidases to yield the active compound that targets the messenger RNA processing pathway in *Trypanosoma cruzi* [7]. Studies in nonhuman primates showed that this compound was uniformly curative of a range of genetically distinct *T. cruzi* lineages with chronic naturally acquired infections. There was no acute or long-term adverse effects when administered orally for 60 days in 19 macaques. This is the first highly effective drug for *T. cruzi* discovered in >50 years. Future clinical studies in humans are eagerly awaited.

### 17.6.1 New Antimalarial Agents in Development

Emergence and spread of artemisinin-resistant *Plasmodium falciparum* and emerging chloroquine resistant *Plasmodium vivax* are of major concern to global health and requires urgent action. Fortunately, in the last 10 years several pharmaceutical companies have developed new antimalarial drugs, and there are 14 active compounds in phase 1 and 2 studies [8]. Several of these agents have promising pharmacokinetic and pharmacodynamics (PK-PD) properties with terminal half-lives of 14.7 to 483 hours and parasite clearance half-life of 3.4 to 9.4 hours with a single-dose monotherapy.

One of the leading candidates for full development is artefenomel, a synthetic trioxolane endoperoxide, which has multiple mechanisms of action and prolonged elimination half-life of 46–62 hours, allowing for single dose therapy in combination with other drugs for *P. falciparum* and *P. vivax* malaria [9]. Phase 2 studies have been performed in combination with piperaquine and ferroquine, and a phase 3b trial in combination with piperaquine is underway [10]. Ferroquine is a ferrocenyl derivative of chloroquine that is active against chloroquine-resistant *P. falciparum* strains [11].

Another promising new antimalarial agent is ganaplacide, an imidazolopiperazine class, with activity against *P. vivax* and *P. falciparum.* This chemical family inhibits *Plasmodium* P1[4]K activity and appears to have activity against multiple stages of the plasmodium lifecycle [12]. The results of a phase 2b study of ganaplacide/lumefantrine versus artemether/lumefantrine showed similar rates of clearance and median parasite-clearing times in children with falciparum malaria (unpublished, News release by Novartis, Sep.29, 2021).

Cipargamin, which targets a new molecular site for malaria therapy (a cell membrane channel in the parasite), is a potent antimalarial in phase 2 studies. In a recent phase 2 dose-escalating study in uncomplicated falciparum malaria in sub-Saharan Africa, cipargamin monotherapy as single doses of 50 mg to 150 mg was compared to artemether/lumefantrine as control [13]. At single doses of cipargamin rapid parasite clearance (median of 8 hours versus 24 hours for the control) occurred with parasitic response rate of >65% at 28 days.

## 17.7 New Antivirals in Development

There are multiple new drugs and repurpose drugs being studied for COVID-19 and some for combination therapy [14]. Many of these are monoclonal antibodies, anti-inflammatory and biological agents. One of the promising antiviral agents, tempol, suppressed the activity of the viral RNA replicase and was in a phase 2/3 trial for high-risk subjects with COVID-19 for outpatient treatment versus placebo. The trial was recently halted early due to lack of efficacy (Adamis Pharmaceutical Corporation news release Sept. 21,2022). This may be due to infection with a less virulent omicron variant, as the hospitalization rate in this trial (<1%) was lower than previous COVID-19 treatment trials. This may pose a logistical problem for future clinical trials with other agents, unless more virulent variants reappear.

The pipeline for new ART in development is fostered by three research-based drug companies, Gilead Sciences, Merck/MSD, and ViiV Healthcare, and all are focused on simplified dual regimens. These involve long-acting compounds of new drug classes and greater potency (HIV i-Base: Pipeline report 2021: HIV drugs in development; 17 Sept. 2021).

### *17.7.1 New Antivirals for HIV*

Lenacapavir is a first-in-class capsid inhibitor with a multi-stage mechanism of action with no known resistance to other existing drug class, and its prolonged half-life allows for subcutaneous dosing every 6 months. In a recent phase 3 trial, lenacapavir with failing optimized background therapy was more effective than placebo in reducing viral load at 15 days in patients with MDR-resistant HIV infection and maintained viral load <50 copies at 26 weeks in 81–83% [15]. The drug is being investigated for treatment-naïve and MDR treatment and preexposure prophylaxis (PrEP) given every 6 months. No serious adverse events have been noted so far.

Islatravir is a first-in-class nucleoside reverse transcriptase translocation inhibitor with multiple mechanisms of action, long half-life with potential for long-acting HIV therapy and prevention. Preliminary results with a fixed combination with doravirine daily showed good tolerance and efficacy [16]. Islatravir was paired with an experimental long-acting nonnucleoside reverse transcriptase inhibitor (NNRI) MK-8507 in a once weekly regimen in a phase 2 IMAGINE-DR trial, but the FDA put a hold on this trial and others with this drug for long-acting treatment in December 2021 (Highleyman L; aidsmap: 21 Dec. 2021). This was related to preliminary results of decline in CD4 counts in patients receiving the long-acting combination.

Recently, Merck announced initiating a new phase 3 clinical program with once daily doravirine 100 mg/islatravir 0.75 mg for treatment of HIV-1 infection, but the trial of once weekly islatravir/lenacapavir remain on hold (Merck news release, Sept. 20, 2022).

Two second-generation maturation inhibitors, GSK3640254 and GSK373937, are undergoing phase 1 and phase 2a studies. Maturation inhibitors (MI) offer a novel mechanism of action at the late stage of the viral lifecycle by producing non-infectious underdeveloped HIV. Initial maturation inhibitors displayed clinical efficacy but were associated with emergence of resistance and gastrointestinal intolerance. The potentially safer new generation MI display strong antiviral activity [17] with good tolerance with daily oral dosing (GSK3640254), while the long-acting injectable (GSK3739937) is undergoing phase 1 study (Pipeline report 2021: HIV drugs in development/ HTB / i-Base).

A new fusion inhibitor (albuvirtide), blocking HIV attachment to CD4 cells, is in phase 2 studies for MDR-HIV. It is a long-acting drug given by weekly injections shown to be effective in a two-drug regimen with oral ritonavir-boosted lopinavir in a randomized, controlled, phase 3 study [18].

There are at least 12 broadly neutralizing monoclonal antibodies (bNAbs) being studied for HIV treatment, prevention, and cure (Pipeline report 2021: HIV drugs in development I HTB I HIV i-Base). A novel immunotherapy drug (N-803), an IL-15 superagonist, was shown to decrease the number of HIV-infected cells up to 6 months after therapy in 11 HIV subjects on ART for at least one year [19].

### 17.7.2 New Antivirals for Chronic Hepatitis B

Chronic hepatitis B virus (CHBV) infection remain a global problem with 296 million people infected and ~ 820,000 deaths per year from cirrhosis and hepatocellular cancer, but no effective cure available. Currently available therapies rarely lead to functional cure or loss of the surface antigen (HBV-sAg). In patients with CHBV with functional cure, the HBV can reappear with suppression of the immune system, probably due to persistence in the hepatocytes.

There are a few new agents in phase 2–3 studies for CHBV aiming to achieve high rates of functional cure and one for chronic hepatitis D virus (CHDV) infection. Bepirovirsen (administered 1–2 times per week sc. for 4 weeks), an antisense oligonucleotide targeting all HBV messenger RNAs, in a phase 2 study showed promising results with favorable safety profile [20].

Vir Biotechnology has two candidates that may best be used in combination to produce functional cure of CHBV. VIR-2218, the foundational candidate, is a GalNAc-conjugated small interfering ribonucleic acid (siRNA) designed to target HBV's X gene region (administered sc.) is in phase 2 study; and VIR-3434 an Fc-engineered human antibody against HBV-sAg with multiple potential mechanisms of action is in phase 1 study [21]. The two agents are also being assessed in a phase 2 trial of CHDV infection.

Bulevirtide is a first-in-class entry inhibitor for the treatment of CHDV, which showed virological and biochemical responses in 2 phase 2 studies; interim report from a phase 3 trial showed after 48 weeks almost 50% achieved reduced or undetectable HDV RNA levels and normalized liver enzymes (Freeman S, Internal Medicine News, at ILC 2022; July 5, 2022).

## 17.8 Summary of New Antimicrobial Agents in Development

The future for development of new antibacterial agents for clinical use appears to be promising with many new drugs in phase 1–3 studies; however, only a few of the traditional agents are new classes targeting new molecular sites, especially new anti-TB agents. It is surprising that only a few of the agents are being developed for MDR gram-negative bacilli such as CRE and MDR-*P. aeruginosa*, mainly new BLIs. Novel agents for treatment of MDR-*N. gonorrhoeae* (in phase 3 studies) are eagerly awaited. It is also disappointing that there are only six bacteriophage products in development in the nontraditional group of antibacterials, mainly for chronic respiratory infections.

There is optimism for treatment of MDR-fungal infections and rare filamentous fungi with no available treatment with the development of seven new agents. It is encouraging and remarkable that four of these are new classes with novel mechanisms of action. The future for treating refractory fungal infections appears to be bright.

Development of new anti-parasitic agents for neglected tropical disease is still being neglected, except for the evolvement of a novel promising drug for Chagas disease. However, the urgent need to find new treatment for drug-resistant malaria is being actively pursued with several promising agents on the horizon.

Research and development of new antivirals seems to be limited to finding new treatments for COVID-19, HIV infection, and CHBV and CHDV. There does not appear to be any new agents in phase 1–3 studies for therapy of cytomegalovirus (CMV) and respiratory syncytial virus (RSV). Development of promising novel agents for CHBV aiming for high rates of functional cure is very encouraging. It is remarkable that new classes of antivirals with different mechanisms of action from traditional agents for HIV infection are being discovered and developed.

It is predictable that microbes will always develop resistance to new antimicrobials, and we will need to maintain a steady supply of novel agents in the pipelines to overcome the emerging resistance.

## References

1. McKenna M. The antibiotic paradox: why companies can't afford to create life-saving drugs. Naure, News Feature, 19 August 2020.
2. World Health Organization. 2021 Antibacterial agents in clinical and preclinical development: an overview and analysis. 2022, Geneva. Licence: CC BY-NC-SA 3.0 IGO
3. Butler MS, Gigante V, Sati H, et al. Analysis of the clinical pipeline of treatment for drug-resistant bacterial infections: despite progress, more action is needed. Antimicrob Agents Chemother. 2022;66:e01991–21.
4. Kaye KS, Belley A, Barth P, Lahlou O, Knechtle P, Motta P, Velixcitat P. Effect of cefepime/enmetazobactam vs piperacillin/tazobactam on clinical cure and microbiological eradication in patients with complicated urinary tract infection or acute pyelonephritis. A randomized clinical trial. JAMA. 2022;328:1304–14.

5. Fisher M, Alastruey-Izquierdo A, Berman J, et al. Tackling the emerging threat of antifungal resistance to human health. Nature Rev Microbiol. 2022;20:557–71.
6. McCarty TP, Pappas PG. Antifungal pipeline. Front Cell Infect Microbial. 2021;11:732223. https://doi.org/10.3389/fcim.2021.732223.
7. Padilla AM, Wang W, Akama T, et al. Discovery of an orally active benzoxaborole prodrug effective in the treatment of Chagas disease in non-human primates. Nature Microbial. 2022;7:1536–46.
8. Abd-Rahman AN, Zaloumis S, McCarthy JS, Simpson JA, Commons RJ. Scoping review of antimalarial drug candidates in phase 1 and II development. Antimicrob Agents Chemother. 2022;66:e01659–21.
9. Phyo AP, Jittamala P, Nasten FH, et al. Antimalarial activity of artefenomel [0Z439], a novel synthetic antimalarial endoperoxide, in patients with *Plasmodium falciparum* and *Plasmodium vivax* malaria: an open-label phase 2 trial. Lancet Infect Dis. 2015;16(1):61–9. https://doi.org/10.1016/S1473-3099[15]00320-5.
10. Belete TM. Recent progress in the development of new antimalarial drugs with novel targets. Drug Des Devel Ther. 2020;14:3875–89.
11. Salim M, Khan J, Ramirez G, Murshed M, et al. Impact of ferroquine on the solubilization of artefenomel [OZ439] during in vitro lipolysis in milk and implications for oral combination therapy for malaria. Mol Pharm. 2019;16:1658–68.
12. Gamo FJ. Drug resistance. 2014, In drug discovery today: technologies. Science Direct Topics.
13. Schmitt EK, Ndaysisaba G, Yeka A, et al. Efficacy of cipargamin [KAE609] in a randomized, phase II dose-escalating study in adults in Sub-Saharan Africa with uncomplicated *Plasmodium falciparum* malaria. Clin Infect Dis. 2022;74:1831–9.
14. Ledford H. A deluge of new drugs for COVID. Nature. 2022;603:25–7.
15. Segal-Maurer S, Dejesus E, Stellbrink H-J, et al. Capsid inhibition with lenacapavir in multidrug-resistant HIV-1 infection. N Engl J Med. 2022;386:1793–803.
16. Molina J-M, Yazdanpanah Y, Saud AA, et al. Brief report: efficacy and safety of oral islatravir once daily in combination with doravirine through 96 weeks for treatment-naïve adults with HIV-1 infection receiving initial treatment with islatravir, doravirine, and lamivudine. J Acquire Immune Defic Syndr. 2022;91:68–72.
17. Dicker I, Jeffrey JI, Protack T, et al. GSK3640254 is a novel HIV-1 maturation inhibitor with an optimized virologic profile. Antimicrob Agents Cheomther. 2022;66:e0187621.
18. Su B, Yao C, Zhao Q-X, Cai W-P. Efficacy and safety of the long-acting fusion inhibitor albuvirtide in antiretroviral-experience adults with human immunodeficiency virus-1: interim analysis of the randomized, controlled, phase 3, non-inferiority TALENT study. Chinese Med J [Engl]. 2020;133:2919–27.
19. Miller JS, Davis ZB, Helgeson E, et al. Safety and virologic impact of the IL-15 superagonist N-803 in people living with HIV: a phase 1 trial. Nat Med. 2022;28:392. https://doi.org/10.1038/s41591-021-01651-9.
20. Yuen M-F, Heo J, Jang J-W, et al. Safety, tolerability and antiviral activity of the antisense oligonucleotide bepirovirsen in patients with chronic hepatitis B: a phase 2 randomized controlled trial. Nat Med. 2021;27:1725–34.
21. Hwang C, Pang P. Addressing the unmet need for a functional cure for hepatitis B. natrureportfolio. www.nature.com/articles/d42473-022-00083-4

# Index





I. W. Fong, *New Antimicrobials: For the Present and the Future*, Emerging Infectious Diseases of the 21st Century,
https://doi.org/10.1007/978-3-031-26078-0

MIX
Papier aus verantwortungsvollen Quellen
Paper from responsible sources
FSC® C105338

If you have any concerns about our products,
you can contact us on
**ProductSafety@springernature.com**

In case Publisher is established outside the EU,
the EU authorized representative is:
**Springer Nature Customer Service Center GmbH**
**Europaplatz 3, 69115 Heidelberg, Germany**

Printed by Libri Plureos GmbH
in Hamburg, Germany